# Genetically Modified Food

# Genetically Modified Food

## A Short Guide for the Confused

ANDY REES

Pluto Press

LONDON • ANN ARBOR, MI

First published 2006 by Pluto Press
345 Archway Road, London N6 5AA
and 839 Greene Street, Ann Arbor, MI 48106

www.plutobooks.com

British Library Cataloguing in Publication Data
A catalogue record for this book is available from the British Library

ISBN    0 7453 2440 1 hardback
ISBN    0 7453 2439 8 paperback

Library of Congress Cataloging in Publication Data applied for

10   9   8   7   6   5   4   3   2   1

Designed and produced for Pluto Press by
Chase Publishing Services Ltd, Fortescue, Sidmouth, EX10 9QG, England
Typeset from disk by Stanford DTP Services, Northampton, England
Printed and bound in the European Union by
Antony Rowe Ltd, Chippenham and Eastbourne, England

*To My Parents*

# Contents

# Acknowledgements

Thanks very much to Dr Arpad Pusztai, Dr Susan Bardocz, Dr Stanley Ewen, and particularly to Dr Michael Antoniou who very kindly gave of their precious time to read and check the manuscript; also to Ben Ayliffe of Greenpeace, Jo Edwards of Transport 2000, Jochen Koester, Robert Parker of Amnesty International, author Jeffrey M. Smith, Brian Tokar, and Robert Vint of Genetic Food Alert UK for supplying information.

I'm also very grateful to the following friends, who read the draft manuscript and gave their valuable opinions: Helen de Castres, Carolyne Hooper, Jane Leavey, Cathy Macbeth, Chris Park, and above all Claire Robinson, who in addition helped me in many other ways.

A big thank you to all the authors and journalists whose work I have collated in order to make this book, and without whom the book would never have been written.

I take my hat off to Jonathan Matthews for his ceaseless and pioneering work at GM Watch, which is an amazing resource that eases the workload of anti-GM campaigners around the world.

And finally, to the great unsung hero, Maharishi Mahesh Yogi, who issued a call to arms to the Natural Law Party to oppose Genetic Engineering before anyone else had blown the whistle.

# Foreword

*Dr Michael Antoniou*

The advocates of GM crops promised much at the time of their launch in 1996. We were reassured that GM was simply a natural extension of traditional plant breeding methods, only more precise and safer; GM herbicide-resistant crops would only need one application of herbicide during a growing season; GM crops engineered to produce their own pesticide would not need spraying with any additional pesticide at all; crop management would be far easier and yields would increase as would profits for farmers. And this was only the beginning. Future developments would include nutritionally 'enhancing' our food crops.

Now, a mere ten years after GM crops were introduced and these bold assertions made, it is clear that none of these promises have been met. Farmers found from the outset that they needed to apply herbicide anywhere between two and five times within a growing season and spray additional pesticide for adequate protection of GM crops. Rather than being easier to manage, GM crops have proved to be a nightmare as weeds and insect pests became resistant to the chemicals either used on or engineered into GM crops. Overall yields have at best been no better than with non-GM conventional varieties and in the case of GM soya have been consistently lower. Problems with GM crop performance have been common, with the latest being the GM cotton failures in India, which led the state of Andhra Pradesh to ban the growing of this crop within this region.

The failure of GM crops to live up to their expectations will come as no surprise to anyone who knows and understands anything about modern concepts of molecular biology. GM technology is based upon a grossly outmoded understanding that genes function as isolated units of information that can be moved around between totally unrelated species with precision and totally predictable outcomes. We now know that this is simply not the case. Gene organisation within DNA is non-random and highly structured, with a clear tendency for genes possessing related functions to be clustered together, which assists in their coordinated control and balanced function. It is therefore no surprise that the GM process, which relies on the random insertion of an artificial gene construct from one organism into usually a totally unrelated one, for example the transfer of a virus, bacteria, unrelated plant or animal or human gene into a food crop, results in major disruptions in the function of host genes as well as the introduced GM gene. In addition, the GM process in general, through unknown mechanisms, turns out to be highly mutagenic; that is, it causes hundreds or even thousands of alterations throughout the host plant DNA, with potentially far-reaching and damaging consequences. As a result it is not surprising to find that there have now been numerous laboratory studies showing potential health-damaging effects from the consumption of GM foods, which have been systematically denied and/or ridiculed by the authorities entrusted with the well-being of people and the environment rather than followed up with further research to ascertain the cause of the problem.

Why is it that despite these problems, GM crops have had such a large uptake albeit in a limited part of the world? The only way a flawed technology that fails to meet its expectations such as GM crops can be brought to and maintained in the market is artificially, through manipulation and what many would call deception (promoted as something other than what it really is), by powerful corporate and political factors. Andy Rees's book brilliantly sets out the first totally comprehensive, well-referenced account of the events that have taken place in the years since the appearance of GM crops on the marketplace. Written in an accessible and engaging manner, it expertly highlights the type of strategic activities carried out by the GM crop biotechnology industry and national governments to push an unnecessary, failed and failing technology on an unwanting world population.

Dr Michael Antoniou, Reader in Molecular Genetics at Kings College, University of London, Guy's Hospital Campus

# Abbreviations and Acronyms

ABC  Agricultural Biotechnology Council (UK)
ACRE  Advisory Committee on Releases to the Environment (UK)
ACNFP  Advisory Committee on Novel Foods and Processes (UK)
ARM  Antibiotic Resistant Marker gene
BBSRC  Biotechnology and Biological Sciences Research Council (UK)
BIO  Biotechnology Industry Organization (US)
BMA  British Medical Association (UK)
Bt  *Bacillus thuringiensis*
CaMV  Cauliflower mosaic virus
CGIAR  Consultative Group on International Agricultural Research
DEFRA  Department of the Environment, Food, and Rural Affairs (UK)
DFID  Department for International Development (UK)
EC  European Commission
EPA  Environmental Protection Agency (US)
EU  European Union
FAO  Food and Agriculture Organisation (UN)
FDA  Food and Drug Administration (US)
FoE  Friends of the Earth (international environmental NGO)
FSA  Food Standards Agency (UK)
FSEs  Farm Scale Evaluations (UK)
GE  Genetic Engineering
GEAC  Genetic Engineering Approval Committee (India)
GM  Genetically Modified
GMOs  Genetically Modified Organisms
GURT  Genetic Use Restriction Technology
ha  Hectare (1 Ha = 2.475 acres; 1 acre = 4,840 sq yds)
HGT  Horizontal Gene Transfer
HT  Herbicide Tolerant
HYV  High-Yielding Variety
IARCs  International Agricultural Research Centres
IMF  International Monetary Fund
ISAAA  International Service for the Acquisition of Agri-Biotech Applications

| | |
|---|---|
| ISIS | Institute of Science in Society |
| JIC | John Innes Centre (UK) |
| MAFF | Ministry of Agriculture Fisheries and Food (UK) |
| MAS | Marker-Assisted Selection |
| NFU | National Farmer's Union (UK) |
| NGO | Non Governmental Organisation |
| OECD | Organization for Economic Co-operation and Development |
| PR | Public Relations |
| rBGH | Monsanto's recombinant (GM) Bovine Growth Hormone |
| rBST | recombinant bovine somatotropin (synonymous with rBGH) |
| RR | Roundup Ready |
| RS | Royal Society (UK) |
| RSPB | Royal Society for the Protection of Birds (UK environmental NGO) |
| SAP | Structural Adjustment Programme |
| SAS | Sense About Science (UK) |
| TNC | Transnational Corporation |
| TRIPS | Trade Related Intellectual Property Rights agreement (WTO) |
| UN | United Nations |
| USAID | US Agency for International Development |
| USDA | US Department of Agriculture |
| WFP | World Food Programme (UN) |
| WHO | World Health Organization (UN) |
| WTO | World Trade Organization |
| WWF | World Wide Fund for Nature Conservation (international environmental NGO) |

# 1
# Introduction

'Who the hell is going to be in charge? A handful of corporate greed-heads, or we the people?' That's what it comes down to. Who's going to be making the decisions in a society that supposedly is self-governing?' – Jim Hightower, writer, editor, and former Texas Agricultural Commissioner[1]

This is a book that shouldn't need to be written. If we lived in a sane country, at a sane period of time, what I am about to write would all be commonsense.

I should need only to write that dousing 97 per cent of our food with highly toxic chemicals is stupidity beyond belief; and that it is inconceivable that we should then want to genetically tinker the food that has served us so well for tens of thousands of years, when the only beneficiaries will be the multinational corporations who invented this so-called technology. End of book.

But sadly, we live at a time when the all-pervading Big Business lobby, and its huge influence on the media, has shifted received wisdom far away from commonsense and brainwashed us with half-truths, bad science, and outright lies. Nowhere have they done this more effectively than the GM debate, where the biotech lobby has poured colossal resources into its aggressive and proactive PR machine. Those scientists who have the temerity to oppose GM crops are set upon with insults and vitriol, often losing their reputations, sometimes their jobs.

This book, therefore, is written in large part to counter this campaign of misinformation. But it is also much more than that. If Big Business is up to all these dirty tricks in the GM arena, it is almost certainly doing the same thing in all areas of our lives. Therefore, I would like to propose a more radical hypothesis, which is that the world is run by the rich and powerful for the benefit of the rich and powerful – essentially an amalgam of state and corporate power. The rest of us are useful only to the extent that we create and consume their products.

Only when all the inconsistencies of modern life are viewed from this angle do the horrors of the world begin to make any sense. When we understand that what suits Big Business is generally what goes, everything then starts to slot into place. It explains why apparently simple and commonsense decisions like disallowing GMOs (Genetically Modified Organisms) are made so complicated. Or why the net destruction of the environment grows each year and yet we do nothing, because we are told 'we need more studies'. Or why the growing cancer epidemic, which kills 40 per cent of Britons, is never discussed in terms of being a disease of industrial chemicals and pollutants.[2] Or why the US spends $360 billion a year on 'defence',[3] when more than 25 million Americans depend on charities to eat[4] and food shortages affect 800 million in the poorest countries.[5] Or why SARS received endless coverage and killed only 774 people in 2003,[6] yet AIDS killed 2.4 million Africans in 2002 alone and is largely ignored.[7] Or why we have endless media coverage of plane or train crashes, murders, sport, and celebrity prittle-prattle, but virtually nothing on the really big issues.

The hypothesis I am proposing will, however, require that we let go of the comforting notion that those in power are essentially acting in our best interests – give or take a few colourful indiscretions here and there – rather than doing what is best for a small, powerful elite. Or as leading American playwright Arthur Miller once said: 'Few of us can easily surrender our belief that society must somehow make sense. The thought that the State has lost its mind and is punishing so many innocent people is intolerable. And so the evidence has to be internally denied.'[8]

This book gives enough evidence, I believe, to put this hypothesis on a pretty strong footing. And, hopefully, it might in some small way help to shift some of that internal denial and allow the average person to make more sense of this ever more nonsensical world we live in.

The struggle between people and corporations will be the defining battle of the twenty-first century. If the corporations win, liberal democracy will come to an end.[9] – George Monbiot, author, environmentalist, and Visiting Professor at the Department of Environmental Science, University of East London

# 2
# An Overview

Genetic engineering is the biological equivalent of splitting the atom and has equally, if not greater, hazardous consequences for humankind. – Dr Robert Anderson, Member of the Physicians and Scientists for Responsible Genetics (New Zealand)[1]

Within ten years, we will have a moderate to large-scale ecological or economic catastrophe [from the use of GMOs]. – Professor Norman Ellstrand, ecological geneticist at the University of California[2]

## SETTING THE SCENE

Firstly, let's get one thing straight: the Genetic Engineering debate can be boiled down to a single sentence. We risk potential health, environmental, and agronomic calamities just so that a handful of corporations can sate their voracious appetite for profit by patenting the seeds of the very food we eat, and then go on to control the global food chain. All else is froth and distraction.

Secondly, the term Genetic Engineering (GE) is an insult to the noble profession of engineering. Engineering implies accuracy, and if there is one characteristic trait about GE – and its proponents, for that matter – it is a complete lack of the virtue. Its proponents would like us to believe that GE is simply a case of carefully removing the desired gene from its source and inserting it into just the right place in the recipient genome (an organism's DNA). But Dr Arpad Pusztai, a scientist with 35 years of lab experience and the world's leading expert on plant lectins (plant proteins that are central to the GM controversy), doesn't agree:

Think of William Tell, shooting an arrow at a target. Now put a blindfold on the man doing the shooting and that's the reality of the genetic engineer when he's doing a gene insertion. He has no idea where the transgene [a GM gene inserted into a new host] will land in the recipient genome.[3]

So, when I use the term Genetic Engineering or Genetically Modified (GM) or technology, think GT – Genetic Tinkering.

## Out-of-date science

According to Dr Barry Commoner (and many other scientists): 'The biotechnology industry is based on science that is forty years old and conveniently devoid of more recent results.'[4] 'The fact that one gene can give rise to multiple proteins ... destroys the theoretical foundation of ... [this] industry.'[5]

Dr Michael Antoniou, Reader in Molecular Genetics at Kings College, London, Guy's Hospital Campus, adds to this:

But what really makes GE out of date is the fact that it is based on the idea that genes are isolated units of information that can be moved around and still have the same effect. This is early 1980s thinking. Now we know that genes work as part of complex interconnected networks and that no gene works in isolation of other genes. When you insert a new gene, the function of both the transgene and the host's genome are disrupted. If you take a gene out of context, you can't predict the outcome and you will not have complete control over the result. This is the key conceptual flaw with GE. This is why GE will never work. It is based on flawed science.[6]

### A BRIEF RECENT HISTORY

Let's go back to the beginning, to where it all started.

### The history of industrial agriculture

The 1940s saw the birth of industrial agriculture in the developed countries, where it is now ubiquitous – 96 per cent of farmland is industrially farmed in Britain.[7] Industrial (otherwise known as chemical, conventional, or high-input) agriculture is the practice of farming with artificial fertilisers and agrochemicals, such as herbicides, fungicides, and pesticides. Agrochemical usage began to take off in the 1950s, peaked in the late 1980s, and is still used in massive amounts to this day.[8] This new form of agriculture appeared to be a great success at first, with record yields in the developed countries in the 1950s and in the developing countries in the early 1960s.

Industrial agriculture was introduced to the developing countries in the 1960s, with the ostensible aim of increasing food production. This was known as the Green Revolution, and it too was based around high-yielding varieties (HYVs) that relied on high inputs of fertilisers, pesticides, and often irrigation. These HYVs were developed by the International Agricultural Research Centres (IARCs) – which were set up in the developing countries in the early 1960s – and promoted to small-scale and subsistence farmers by the International Monetary Fund (IMF), the World Bank, and the IARCS.[9]

Yields did increase for a while, but then so did a long list of problems that accompanied the Green Revolution. Far from helping the poor, it concentrated wealth and land into the hands of the wealthier farmers, who could afford the expensive inputs, and drove thousands of subsistence farmers into debt or off the land. The steep increase in agrochemicals led to a huge loss of natural biodiversity and has poisoned farmers, groundwater, land, and the environment. Irrigation has depleted water resources, and increased salinity, often leaving land unusable. Nutrient levels in the soil and crops dropped, instigating new problems of nutrient deficiency. Pests and diseases often got worse, as HYVs were less pest resistant than traditional varieties and relied on monocultures. Yields that at first rose, plateaued, and are now falling. In the process, agriculture was transformed from subsistence farming into agribusiness, so opening it up to agricorporations and their products.[10]

The hunger that the Green Revolution was meant to eradicate has remained stubbornly intransigent, with the number facing hunger and malnutrition worldwide remaining at 800 million.[11]

### Macro-economic policies

A look at world politics will give some hints as to why there is so much need amid so much plenty. It also helps to explain the unprecedented ascendancy of the transnational corporations (TNCs).

In the 1970s, the two OPEC oil price rises were accompanied by a lending spree to poor countries. This money was often invested in transforming subsistence agriculture into chemically farmed tropical cash crops, like bananas and cocoa, for export (resulting in the best land becoming unavailable to feed local populations). There then followed a mix of a fall in commodity prices, a recession in the developed countries where these commodities were bought, and increasing interest rates, which led to the long-term debt of the developing countries soaring. With massive defaults in the offing, the World Bank and IMF reacted by imposing Structural Adjustment Programmes (SAPs). This meant stringent cutbacks on education and health spending, a reduction in workforces (that is, mass unemployment), the deregulation of environmental controls, and a heavy emphasis on export earnings (with yet more cash cropping). The financial institutions used this situation to dictate advantageous entry terms for foreign corporations in the developing world, thus opening up national industries and natural resources for plunder.

Many believe this was an intentional policy by the rich countries. Certainly, it has increased the power of the biotech corporations. Meanwhile, though, most developing countries have been unable to pay off their debts, and hunger and famine still abound.

Some debt cancellation was agreed in Cologne in 1999, but as of March 2005, it was still only about 10 per cent of that owed by the world's poorest countries.[12] In September 2004, the UK government announced it would help 21 poor countries with debt service relief on money owed to the World Bank and African Development Bank. However, while encouraged by this step, critics want to see debt actually cancelled, and they want it to be applied to all poor countries, without the stringent economic requirements that are usually attached.[13] Furthermore, they also want to see more and better aid plus justice in world trade, which currently robs poor countries of £1.3 billion a day – 14 times what they get in aid.[14]

SAPs are a part of the ideology of free trade, which has been on the increase since the 1980s, promoted by corporations in the World Trade Organization (WTO), with the backing of many developed countries. Under free trade, state intervention in the economy is discouraged, particularly measures that protect industry; yet the US and the European Union (EU) carry on subsidising their farmers, to the point that developed world farmers are unable to compete (another cause of poverty). National and international regulations and treaties protecting the workforce and the environment are often contested through the WTO – which has legislative and judicial powers over sovereign nations – as barriers to trade, thus expanding the rights of corporations. Free trade purportedly spurs competition between corporations, and yet these companies run near monopolies around the world, with 90 per cent of the export market for wheat, corn, coffee, tea, pineapple, cotton, tobacco, jute, and forest products controlled by five companies or less.[15]

In 1994, the US government managed to move the matter of intellectual property from the World Intellectual Property Organization to the WTO, thus making protection of patents on micro-organisms obligatory. This was a real coup, providing the first global mechanism for patents on living beings, and laying the foundation for the gene revolution and the patenting of crops.[16]

### The corporate takeover of the food chain, part 1

As with most areas of business, a handful of corporations now dominate the international food chain, with over 60 per cent of it controlled by just ten companies, which are involved in seeds, fertilisers, pesticides, processing, and shipments. For example:

- Cargill and Archer Daniel Midland control 80 per cent of the world's grain;[17]
- Syngenta, DuPont, Monsanto, and Aventis account for two-thirds of the global agrochemical market.[18]

By the early 1990s, almost all the major product sectors of European food were controlled by two or three companies. For example:

- the top three biscuit suppliers accounted for 70 per cent of the market;
- the top three suppliers of breakfast cereals for 64 per cent;
- the top three manufacturers of snack foods for 91 per cent.

And in the UK:

- just eight processors accounted for 60 per cent of the entire UK food market in 1994;[19]
- the biggest five supermarket chains now sell almost 75 per cent of groceries in Britain.[20]

The agbiotech industry began to consolidate in the 1990s. It had become more and more expensive for the agrochemical corporations to create new pesticides; furthermore, some of its main products would soon come off patent. Biotechnology opened up a vast new arena for profit – that is, the patenting of new or changed genes. Whereas a new pesticide takes $40–100 million to bring to market, it only costs $1 million to market a new plant variety.[21] Hence, these corporations went on a spending spree of small seed companies in the 1990s,[22] so that by 2001, only four corporations sold practically all GM seeds, with a staggering 91 per cent sold by Monsanto alone.[23] We can expect a further consolidation of the non-GM seed market as biotechnology allows the agbiotech corporations the means and incentive to close in on crops that were previously unprofitable.[24]

### The Gene Revolution

This brings us to the so-called Gene Revolution – the cosy label given to genetic engineering in agriculture – which, like the Green Revolution before it, is being heralded as the saviour of the poor. And like the Green Revolution, it too relies on a very one-dimensional approach to hunger – the genetic improvement of crops, and the agrochemicals required by the high-yielding varieties. Yet, as we have seen, the Green Revolution has failed the majority of farmers, the environment, and the hungry – although large-scale farmers and, more importantly, the agricorporations have benefited enormously. It failed mainly because of this simplistic approach to dealing with hunger, while ignoring the fundamental tenets of good farming, such as water management, mixed cropping, soil fertility, and other sustainable practices, which are easily able to double yields. Therefore, far from learning from the Green Revolution, the proponents of GE are simply continuing with more of the same. But why? Is it because the real agenda is not the feeding of the poor at all, but the corporate takeover of the food chain? There is certainly a lot to play for – a huge untapped market for their products, with 90 per cent of African farmers still saving their own seeds[25] and large parts of the developing countries still farming organically.

## WHAT IS GENETIC ENGINEERING AND WHO WILL BENEFIT?

Genetic Tinkering is the process of adding a gene or genes (the transgene) to plant or animal DNA (the recipient genome) to confer a desirable trait, for example inserting the genes of an arctic flounder into a tomato to give antifreeze properties,[26] or inserting human genes into fish to increase growth rates.[27]

But, as we are about to discover, this is a technology that no one wants, that no one asked for, and that no one but the biotech companies will benefit from. This is why the biotech lobby has such a vast, ruthless, and well-funded propaganda machine. If they can reinvent our food and slap a patent on it all, they have just created an unimaginably vast new market for themselves.

And to try to convince a suspicious public, they have given us dozens of laudable reasons why the world will benefit from this tinkering. The companies who so enthusiastically produce millions of tonnes of pesticides every year are now telling us that GMOs will help reduce pesticide use. The companies who have so expertly polluted the world with millions of tonnes of toxic chemicals are now telling us that GM will help the environment. The companies who have so nonchalantly used child labour in the developing countries,[28] and exported dangerous pesticides that are banned in the developed countries to the developing countries, are now telling us they really do care about people and that we must have GM to feed the world. The companies who are suing farmers for alleged GM patent infringement are in the same breath telling us that GM will benefit farmers, which is a mighty curious fact, since in Canada alone it has been calculated that GM wheat farmers could lose about $78 million (mainly through lost markets), while biotech giant Monsanto could pocket a tidy $157 million.[29]

The UK Government's 2003 GM 'Public Debate' showed that an overwhelming 86 per cent of people were unhappy with the idea of eating GM food.[30] The public instinctively knows that Big Business has gone way too far with this one. People are revolted by the idea of inserting the genes of a toad into a potato, or fish genes into tomatoes, or scorpion or spider genes into vegetables.[31] Hardly any wonder, then, that the *Daily Mail* has dubbed GM foods 'Frankenfoods'. Malcolm Walker, Chairman and Chief Executive of Iceland Foods, said in 1996:

Millions of ordinary people are very worried about genetically modified foods and I am one of them... With genetically modified foods I believe we have reached the thin edge of the wedge, we are messing with the building blocks of life and it's scary.[32]

But there's more to it than that. It is one thing to destroy the environment – that's one step removed. Messing with the very food we eat is like declaring corporate warfare on humanity.

## DOES ANYONE WANT GM?

No!

If you are statistically challenged, the subheadings will tell you who doesn't want GM foods; if like me, you have a sad fascination for facts and figures, read on.

### Consumers

Consumers around the world want GM food labelled (94 per cent in the EU,[33] 94 per cent in the UK,[34] 92 per cent in the US,[35] 87 per cent in China,[36] over 90 per cent in Canada[37]) and most don't want it at all (80 per cent in the EU,[38] 86 per cent in the UK,[39] 55 per cent in the US,[40] 56 per cent in China,[41] 68 per cent in NZ);[42] 84 per cent of Australians said they worried about eating GM foods[43] and 92 per cent of Canadians were concerned about the long-term risks of them[44].

**Retailers**

- In Britain, all the major supermarket chains refuse to sell GM food (most since 1999); in fact, the British Retail Consortium, which represents 90 per cent of high-street shops, have told Prime Minister Tony Blair they will refuse to stock GM foods even now GM crops are commercialised.[45]
- In Europe, as a whole, almost all the food manufacturing and retail industry has stopped using GMOs,[46] with the vast majority of European (that is, the EU plus Norway and Switzerland) retail chains 'against' GMOs in their own-brand products. The British lead the field.[47] Swiss retailers have agreed never to sell GM food.[48]
- In the US, supermarket chains Trader Joe's, Wild Oats, and Whole Foods have undertaken to remove GM ingredients from their own-brand lines, owing to consumer pressure.[49]
- In Australia, consumer pressure has spurred the three biggest poultry producers – Inghams, Bartter Steggles, and Baiada, which supply 80 per cent of Australia's chicken – to drop GM feed.[50]
- Similarly, in New Zealand, following activist and consumer pressure, KFC has declared it is 'committed' to carry on using GM-free soy as long as it is available.[51]

**Food importers and producers**

- In the UK, the vast majority of food and drink manufacturers said in 2002 there was little chance of them making products containing GM ingredients for at least the next five years.[52] Most removed GM ingredients in 1999, with British Sugar announcing back in 1998 that it would not use any GM sugar beet.[53]
- In Europe, imports of US corn dropped by a colossal 96 per cent in 1999, and soybean sales by $1 billion a year from 1996 to 1999, because the US could not provide non-GM crops.[54] Canadian oilseed rape has suffered a similar fate, with nearly the entire $300–400 million a year of European sales vanishing.[55]
- In the US, GM potatoes were taken off the market when McDonald's, Burger King, McCain's, and Pringles refused them.[56] Furthermore, domestic buyers, including Frito-Lay, Gerber,

Heinz, Seagrams (North America's largest potato processor), and the entire sugar industry want non-GM crops.[57]

- Australian food company Goodman Fielder, Australia's biggest buyer of oilseed rape (canola) oil, has said it will not buy products made from GM rape, because its consumers don't want GM products.[58]
- Japan – the chief importer of most American food exports – told the US that all US wheat would be unwelcome, if Monsanto commercialises GM wheat.[59] Likewise, US soybean exports to Japan will probably diminish as soybean buyers carry on moving to non-GM soy.[60]
- In China, the world's largest food market, 32 food producers, including leading brands, are now officially committed to not selling GM food.[61]
- And in Russia, in 2004, with high consumer concern about GM ingredients, companies like baker MBKK Kolomenskoe are increasingly guaranteeing GM-free ingredients.[62]

### Top chefs

In the late 1990s, more than a hundred top UK chefs and food writers undertook not to use GMOs, and to urge other chefs and restaurants to follow suit. Euro-Toques, an association of top European chefs, ran a similar campaign.[63]

### Farmers

Even farmers are not keen on GM crops – and the more they learn about them the more they are turning against them. For example:

- In Switzerland, farmers have agreed never to produce GM food.[64]
- In the UK, 53 per cent of farmers were against, with only 36 per cent in favour.[65]
- In Australia, 74 per cent said they would not consider growing GM crops, 49 per cent were generally opposed, while only 23 per cent were supportive.[66]
- In Japan, growers are announcing GM-free crop zones to heighten public awareness and to stop the increase of GM crops.[67]
- Even more telling, in Canada, where they have one of the longest track records with GM crops, 87 per cent of farmers said they didn't want GM wheat.[68]

- And in America, the GM heartland – where farmers apparently worship at the altar of GMOs – few are buying into GM technology, beyond soy and maize for animal feed and cotton for fibre, as packers and processors are afraid of jeopardising their markets.[69] GM tomatoes and potatoes never took hold and were removed from the market, and GM sugar beet, flax, and rice, although authorised, were not commercialised (that is, sold to farmers and grown as commercial-scale crops).[70] Furthermore, nearly a quarter of US grain elevators are demanding segregation of GM crops.[71]

- In Argentina, the world's number two GM crop producer, new product authorisations from the government have almost ceased, after a spate of GMO approvals from 1996 to 1998.[72] And the number of farmers buying certified GM seeds has declined significantly in recent years – down from 50 per cent of soy seeds to 20 per cent.[73] 'Our brief history of submission to the world bio-technology giants has been so disastrous that we fervently hope other Latin American nations will take it as an example of what not to do,' says Jorge Eduardo Rulli, one of Argentina's leading agronomists, only six years after the country decided to embrace GM technology.[74]

**The Church**

Although, back in 2002, the Vatican and Italian Archbishop Renato Martino, lobbied by the US biotech lobby, were keen on GM foods,[75] the church in general felt otherwise, with stiff opposition from Catholic clergy especially in the developing countries, for example Brazil, Zambia, Botswana, South Africa, Swaziland, and the Philippines.[76] In the developed world, resistance to GMOs is coming from the church in Canada,[77] the US,[78] Germany,[79] the Church of England (which has refused to grow GM crops on its land[80]), and the Church of Scotland,[81] to mention but a few.

So, all in all, GMOs seem to be as welcome in the kitchen as a dusting of cyanide.

## HOW FAR HAS IT GONE?

GM crops were first grown commercially in 1996, with Calgene manufacturing the FlavrSavr tomato, the first GM food in the US. Since then, the number of GM crops, and the number of countries commercialising them, has gradually increased. However, just as the biotech lobby has told us how popular biotech foods are, and that resistance just comes from a few cranks, they want us to believe that GM crops now dominate world agriculture. The truth is that they are still very marginal and that the biotech companies are struggling financially, in the face of worldwide opposition to GM. The statistics bear this out:

- The global area of GM crops was only 81 million hectares (ha) in 2004,[82] a mere 1.4 per cent of global agricultural area.[83]
- Six countries – America, Argentina, Canada, Brazil, China, Paraguay – accounted for 98 per cent of this, with another ten making up the rest.[84]
- Four crops made up 99 per cent of this area – soy, cotton, oilseed rape (canola), and maize.
- And there are only two major traits on the market – herbicide resistance and insect resistance. Other more useful traits 'are at least ten years away from commercial reality, and most will fail', says GeneEthics Network Director, Bob Phelps.[85]

This, then, is a mighty interesting notion of world domination. In truth, it is a very US-dominated industry, with:

- America growing 59 per cent of all GM crops in the world;[86]
- US biotech giant Monsanto owning the seeds of more than 90 per cent of all GM crops grown in 2001;[87]
- Monsanto having a patent on all GM soy in Europe and all GM cotton in Europe and the US.[88]

Perhaps the biotech lobby has forgotten that 35 countries with 3 billion people (about half the world's population) have some kind of restrictions on GM food, whether that's mandatory labelling (19 countries) or outright bans.[89] The EU, Latin America, most of Asia, and Africa have, at least temporarily, blocked GM products from their markets. But the US government, as we will see, is putting enormous pressure on many of these countries to buy its technology.

## HAVE GMOs BEEN TESTED?

The Big Lie that underpins the whole business of Genetic Trickery – and the one that surprises most people – is that while the industry claims that GM technology has been adequately tested, and that there are no health risks attached to GMOs, as far as human health goes, it has barely been tested at all.

The biotech lobby got around testing by inventing a supreme nonsense called Substantial Equivalence (SE), which claims that chemically speaking the GM and non-GM version of a plant is so similar as to be the same. This begs the question: if GMOs are 'the same as everything else, then how come they have a patent on them?'[90] A piece in the science journal *Nature* said the concept of Substantial Equivalence is 'pseudo-scientific', a 'commercial and political judgment masquerading as if it were scientific', 'created primarily to provide an excuse for not requiring biochemical or toxicological tests'.[91] As we will see later, in Chapter 5, the fraud of SE was more than illustrated during the StarLink fiasco.

Jack Cunningham, the UK cabinet minister with overall responsibility for biotechnology, rather unwittingly gave the game away (in a leaked internal memo, February 1999): 'Why don't we require a pharmaceutical type analysis of the safety of these foods with proper trials?'[92] Henry Miller, who was in charge of biotechnology issues for the US Food and Drug Administration (FDA) from 1979 to 1994, provides the answer:

In this area, the US government agencies [that is the FDA, the US Department of Agriculture (USDA), and the Environmental Protection Agency (EPA)] have done exactly what big agribusiness has asked them to do and *told them* to do (my emphasis).[93]

Dr Arpad Pusztai says of the current arrangements for checking GM food safety in Britain:

The Advisory Committee on Novel Foods and Processes, as the [UK] regulatory authority, has no laboratory of its own. So, therefore, they have to rely on the data they get from the GM companies... Can you really imagine [that] a GM company, or any company for that matter, will disclose data which will question the safety of their product? It's preposterous. The whole thing is a sham.[94]

Former UK Environment Minister Michael Meacher observed: 'There have been no tests. That is an enormous gap and I think a scandalous omission in making the decision about whether or not these [GMOs] are safe to eat.'[95] And by 'no tests' he meant there was not one published, human study that demonstrates the safety of GM food,[96] except the Newcastle Study, which was published (2004) after his statement was made.

There is precious little other health testing either. Most of the testing that the industry cites is either feeding studies (which look at the commercial value as an animal feed) or compositional studies (studies of Substantial Equivalence) or nutrient and toxin studies (which only test for known toxins and not for potential new toxins caused by GE) or allergenicity testing (which lacks reliable methods). All of these are irrelevant, or seriously flawed, in terms of human health concerns.

Gundula Azeez of the Soil Association writes of Professors Pryme and Lembcke's important review of the lack of food safety data on GM foods from July 2003:

It says that there have only been ten published studies of the health effects of GM food/feed [on animals]. ... Over half were done in collaboration with [biotech] companies (fully or partially), and these found no negative effects on body organs. The others were done independently and looked more closely at the effects on the gut lining. Several [4] of these found potentially negative changes which have not been explained.[97]

The review itself observes: 'It is remarkable that these effects have all been observed after feeding for only 10–14 days.'[98]

Azeez continues:

With only ten studies, many of which were inadequate [5], the hypothesis of the biotechnology companies and the FSA [the UK Food Standards Agency] that GM foods are safe has actually not been proven, because the science to prove it simply does not exist. Moreover, the limited available evidence indicates that there could be negative effects.[99]

Over and above the four studies mentioned above, there are a number of other studies showing potentially negative effects from GM foods (see Resources, p. 237). The Newcastle study, already referred to, showed GM DNA transferring to gut bacteria in humans after a single meal.[100] The unpublished FlavrSavr tomato study resulted in lesions and gastritis in rats.[101] Dr Terje Traavik's unpublished research found GM DNA in rat tissues after a single meal, and it was also confirmed to be active in human cells.[102] Monsanto's 90-day study of rats fed MON863 resulted in smaller kidney sizes and a raised white blood cell count. And Dr Ewen and Dr Pusztai's 1999 ten-day study on male rats fed on GM potatoes (which is one of the four studies mentioned above) showed that feeding GM potatoes to rats led to many abnormalities, including: gut lesions;[103] damaged immune systems;

less developed brains, livers, and testicles; enlarged tissues, including the pancreas and intestines; a proliferation of cells in the stomach and intestines, which may have signalled an increased potential for cancer; and the partial atrophy of the liver in some animals.[104] And this is in an animal that is virtually indestructible. Well, it's good to know what the biotech lobby means by no health risks.

It is not enough for Monsanto et al. to say no proof of danger exists. 'No evidence' is not the same as 'no risk'. According to the Precautionary Principle: 'the proponent of an activity, rather than the public, should bear the burden of proof'.[105]

As Dr Suzanne Wuerthele, a former EPA toxicologist, put it:

This technology is being promoted, in the face of concerns by respectable scientists and in the face of data to the contrary, by the very agencies [the USDA, the EPA, and the FDA] which are supposed to be protecting human health and the environment. The bottom line in my view is that we are confronted with the most powerful technology the world has ever known, and it is being rapidly deployed with almost no thought whatsoever to its consequences.[106]

### ARE GMOs WELL REGULATED?

GM foods have never been properly tested and therefore should never have been approved.

Steven Druker, JD, a US public interest lawyer, and authority on the legal and scientific issues of GM foods, said:

It's astounding that this venture [GM crops] to radically transform the production of our food continues to be portrayed as based in sound science and rigorous regulation when in fact it depends on flagrant disregard of both science and the food safety laws. Such deception ranks as one of the greatest frauds ever perpetrated.[107]

As proof of this lack of effective regulation, we have to look no further than our own (EU) grocery shelves, where – despite the de facto EU moratorium on GM foods from 1998 to 2004 – thousands of products probably contain GM ingredients or GM material. We don't know how many exactly because of loopholes in the labelling regulations, which suit industry very well – what we don't know about, we can't protest against.

The sources of GM contamination are as follows:

- Most is from GM-fed animals. The April 2004 EU labelling laws omitted meat, eggs, fish, and dairy products produced from GM-fed animals. This means that we have no way of knowing which of these products contain GM material or not, unless they are organic. What we do know is that most of the animals that supply non-organic animal, fish, and dairy products in the UK are GM-fed; these end up in vast numbers of processed foods.
- Some is from permissible levels of unlabelled GM contamination of food, animal feed, and seeds. And although these levels have been slightly lowered by the April 2004 EU labelling laws, all this has done is to make such contamination reassuringly official.
- Some is from GM yeasts and enzymes, and additives containing GMOs, which also do not have to be labelled. This is a grey area still being sorted out by the European Commission.
- In the future, a possible 0.3–0.7 per cent contamination level of seed *might* be allowed (before labelling is required),[108] which will result in massive contamination of non-GM crops.
- And then there's the huge contamination of non-GM crops by cross pollination in GM-growing countries (see p. 60).

The bad news is that all of this means that even well-informed people think they are eating a GM-free diet, when they are not. Before long – unless we fight even harder to get our labelling laws tightened and GM thwarted at every opportunity – GM will be in everything. The good news in Europe, though, is that relative to most other countries, EU GM regulations are fairly decent and GM contamination has been kept to a comparative minimum. Consumer rejection and (on the whole) strong anti-GM policies from the supermarkets have seen to this.

In non-EU countries, there is a hotchpotch of quickly drafted regulations, but no country is free from the contamination of their foodstuffs – especially if you live in the US, where people have been eating increasing amounts of GM foods since they were first grown commercially in 1996. According to the Grocery Manufacturers of America, an estimated 70–75 per cent of products on US grocery shelves may contain GM ingredients.[109] Furthermore, there are no laws covering GM, not even the requirement to label.

Once again, the commercial interests of a tiny minority take precedence over the safety of the world's population.

## EU GM legislation

In the EU, where at least a reasonable degree of sanity prevails in the GM department, a fair amount of GM legislation has hit the statute books.

### GM moratorium

No new GM crops or foods were approved for use in the EU between October 1998 and May 2004, when Syngenta's Bt11 maize was approved. In October 1998, EU governments imposed a voluntary moratorium in the face of rising public anger at the way the technology was introduced, without consumers knowing about it. In the UK, GM tomatoes, soy, and corn were quietly approved for human consumption between 1996 and 1998, and, although they were labelled, there was so little publicity that the general public were unaware of what GM was, and what risks it carried. Effectively, people were being used as guinea pigs.[110]

### Labelling rules

New strengthened EU labelling rules on traceability, animal feeds, derivatives and GM thresholds were passed in April 2004. These will:

- strengthen traceability, ensuring that GMOs can be traced 'from farm to fork' and removed from the food chain if problems emerge. Britain, rarely more efficient than when opposing good ideas, voted against the rules;
- extend labelling regulations to include animal feed, for the first time, making it easier for food manufacturers to obtain milk, eggs, and meat from animals fed on GM-free diets;
- extend and strengthen regulations on the mandatory labelling of food products derived from GM crops (derivatives), making it easier for consumers to avoid food containing GM ingredients;
- lower the GM threshold (the amount of GM present in a food product before GM labelling regulations apply) from 1 per cent to 0.9 per cent on all food and feed.[111]

### GM-fed animal produce

On the other hand, meat, eggs, and dairy products from GM-fed animals *still* do not have to be labelled. This is a major omission, given that British farmers alone import 1 million tonnes-plus of

GM maize and soy as livestock feed every year.[112] This is the biggest source of GM food contamination across the EU, where similar things are happening.[113]

Many GM ingredients will also be hidden in processed foods, an enormous number of which contain dairy products and eggs. It is also a huge market for GM products, which means that consumers are still funding the growing of GM crops in places like Argentina, where it has been disastrous for the environment, people's health, and the social fabric of the country.

Research from Germany, kept secret for three years, but uncovered by Greenpeace in June 2004, reveals another problem that the GM lobby (as well as the UK Food Standards Agency) assured us would not happen, namely that GM DNA was found in the milk of GM-fed animals.[114] And even if the amounts of GM material getting through to the milk are small, it is useful to heed Dr Arpad Pusztai's words: '... consuming even minor constituents with high biological activity may have major effects on the gut and body's metabolism'.[115]

The jury is out on how this will affect human health, because so little peer-reviewed research has been done. However, given the record of GM, this puts a serious question mark over the safety of milk (and other dairy and animal products) derived from GM-fed animals.

### GM Seed Contamination Directive

One of the chief causes of GM contamination of the food supply is through seed contamination. Yet, the EU is considering setting the threshold of GM seed contamination at 0.3–0.7 per cent, which may sound small, but translates into 1 in 200 maize seeds. This is effectively commercial-scale release by stealth. It could result in GM contamination on 10 per cent of EU arable land within a year, allowing the release of about 7,000 million unmonitored and unregulated GM rape and maize plants per year. Environmental groups are calling for zero tolerance on GM contamination of seeds above the limit of detection (currently 0.01 per cent). Austria's zero-tolerance policy has worked successfully.[116]

### International agreements

### Cartagena Protocol on Biosafety

The Cartagena Protocol on Biosafety came into force in September 2003, aiming to ensure that GMOs do not harm human health, contaminate traditional crops, or reduce natural biodiversity. It permits the application of the Precautionary Principle by sovereign nations, whereby countries are entitled to fully documented traceability on GM imports. It also means that they can ban crops that might threaten the environment or human health. Bolivia and Croatia have already banned all GMOs.[117]

Although signed by over a hundred countries, the US has opposed the protocol,[118] once again defying the international order, as it has done with the Kyoto Protocol, the Biological Weapons Convention, the International Criminal Court, and the invasion of Iraq in 2003, to name only a few. Nevertheless, in a breathtaking display of arrogance, the US still sent about a hundred lobbyists to the UN conference on biodiversity in Kuala Lumpur in February 2004, where they tried to scupper the deal, even though the US, as a non-signatory, wasn't officially allowed to take part in the process.[119]

### Trade Related Intellectual Property Rights (TRIPS) agreement

The World Trade Organization's (WTO) TRIPS agreement came into force in 1995 and is compulsory for all WTO member countries. It has massively eroded the rights of the developing countries, and most believe it will result in the theft of developing countries' resources by the biotech companies, because it has effectively globalised the patent system, granting corporations the right to protect their invention (that is GM plant seeds) in all WTO member countries for 20 years.

This gives a strong incentive for large corporations to appropriate (pirate) and patent seed varieties that local farmers have taken many generations to develop. In other words, seed corporations will monopolise and sell what until now has been free – the right to save seed from one year until the next – turning it into an expensive commodity. This will destroy crop diversity and erode farmers' rights.

Interestingly, the Intellectual Property Rights Committee was composed of 13 leading US corporations (including Monsanto and DuPont), who basically wrote the TRIPS agreement. The developing countries were strongly opposed to this agreement. But the multinationals – with their ravenous appetite for profit – were robustly in favour. By 2001, they had slapped thousands of patents on food crops, with six corporations controlling 74 per cent of all biotech crop patents.[120] As with all trade issues, the biotech industry's ideal situation would be one global process, giving worldwide cover to each patent, which would speed up their takeover of the food chain immensely.

# 3
# The Players

This chapter will take a look at some of the main players involved in this debate. However, given space restraints, I have singled out only the most active and relevant organisations. No offence is meant to those who feel their behaviour was bad enough to warrant inclusion – I'm sure it was.

## INDUSTRY AND ITS LOBBY GROUPS

### Biotech, agrochemical, and associated companies

#### Monsanto

Monsanto can boast the following character reference – that an Alabama court found it guilty of 'behavior so outrageous in character and extreme in degree as to go beyond all possible bounds of decency so as to be regarded as atrocious and utterly intolerable in civilized society'.[1] Norman Baker, MP, said that Monsanto is 'public enemy number one. They insist on thwarting consumer choice, bulldozing elected governments, and forcing their wretched products on the world's population.'[2]

This all-American corporation has a very dirty history. This includes the manufacture of DDT, a highly toxic insecticide; this was banned from use in the developed countries in 1976, yet it is still sold by Monsanto et al. in many developing countries. The company also manufactured other exceptionally unpleasant chemicals, for example PCBs (Polychlorinated Biphenyls) and Agent Orange, the cause of both massive human suffering and environmental damage.[3] Company documents – many of them labelled 'confidential: read and destroy' – showed that there were problems with PCBs as far back as 1938. The Environmental Protection Agency (EPA) was aware of the problem from the 1970s, but didn't warn the residents of Anniston, Alabama, where the chemical was manufactured and where it has wreaked havoc on people's health.[4]

Monsanto is also powerful and influential. It is the world's second largest agrochemical corporation,[5] and the second largest seed company.[6] In 2005, it had global sales of $6.29 billion – 40 per cent from Roundup and other glyphosate products, and 34 per cent from seed and genomics (the analysis of genes and their functions in an organism).[7] In 2002 Roundup was the world's biggest-selling herbicide and its Roundup Ready seeds are the world's best-selling GM crops.[8]

During the 1990s, Monsanto invested heavily in biotech research, and spent almost $10 billion globally buying up seed companies.[9] In the mid-1990s, it began pushing aggressively for the commercialisation of GM crops, although having conducted UK field trials since 1995, it took a secondary role to Bayer in the UK GM commercialisation process. Monsanto, and others, are now pressurising the EU to drop its moratorium by flooding it with new applications for GM crops. And in the US and Canada, it was pressing hard for GM wheat commercialisation.[10]

### Syngenta

Syngenta, a British/Swiss corporation, was formed by the merger in 2000 of Novartis and AstraZeneca. In 2002, Syngenta had sales of around $8.1 billion.[11] Currently, it is the world's largest agrochemical corporation and the third largest seed company.[12] It is generally more low-key than Monsanto, except in its energetic acquisition (pirating) of genetic material (see p. 85). It sells biotech crops in the US, Canada, and Spain and has conducted a few UK GM crop trials, which include research on traitor technology (whereby certain traits are turned on by the spraying of specific chemicals). Both Syngenta and Monsanto are trying for commercialisation of GM sugar beet in the UK. Syngenta is probably the biotech company that has most effectively co-opted the sustainable development agenda and pushed for GM crops with purported benefits for the consumer.[13]

### Bayer

Bayer AG was voted one of the ten worst corporations by Multinational Monitor in 2003.[14] Bayer is a gargantuan German chemical and pharmaceutical manufacturer, with global sales of nearly $30 billion in 2000. In 2002, it bought Aventis's contentious CropScience business, to create Bayer CropScience (BCS), which made it a major player in GM crops.[15] In 2005, BCS had sales of €6 billion.[16] BCS is the sixth largest agrochemical company in the world,[17] and it sells GM crops in the US, Canada, Argentina, and Australia. It has been jostling for GM commercialisation in the UK, Europe, and India, with its seeds used in more than 85 per cent of UK GM crop trials in 2002.[18]

### DuPont/Pioneer Hi-Bred

DuPont, the US chemical company, is the second largest chemical manufacturer in the US and the world's fifth largest agrochemical company; its subsidiary Pioneer Hi-Bred is the world's largest seed company, with sales of more than $1.9 billion in 2000.[19] By the late 1990s, DuPont sold its petroleum interests to become a biotech company, with plans also to go into biopharmaceuticals. DuPont and Pioneer's participation in UK GM crops has been small, but globally the company is very large.[20]

### Advanta

Advanta was the sixth largest seed corporation in the world, with €400 million of sales in 2004. It has been engaged in various UK and EU GM field trials. The company is deeply involved in the regulation of GM crops in the UK, and in the EU it is pressing for a relaxation of GM contamination levels in non-GM crops. This is hardly surprising, given the scandal that erupted in the EU when it was discovered that Advanta's non-GM seed stock was significantly contaminated with GM.[21]

### Dow

In 2001, US company Dow was the world's seventh largest seed company and seventh largest agrochemical company.[22] The company that in the 1960s made napalm a household name – an ingenious chemical weapon, made of jellied gasoline, which stuck to its victims and caused terrible burns – has now apparently discovered ethics. It states on its website: 'At Dow, protecting people and the environment is part of everything we do and every decision we make.'[23]

Dow has long been a champion charmer. In 1999, the watchdog INFACT noted:

Dow's own history includes covering up information about dioxin contamination of Agent Orange ... as well as problems related to its consumer products including silicone breast implants, and the pesticide DBCP which Dow continued to sell abroad even after it was banned in the US because it causes sterility.[24]

### Cargill

Cargill is a huge US food-marketing corporation, supplying over 60 per cent of the world trade market. It operates in over 60 countries, with massive revenues of $71 billion in 2005. Its revenues from coffee sales alone are greater than the Gross Domestic Product (GDP) of any one of the African countries from which it buys coffee.[25]

### Pro-GM lobby groups

Many organisations have appeared around the world over the last decade to make sure that the interests of the biotech industry are cultivated at the heart of government, at international negotiations, and in world trade.[26] There follows a little taster, but for much more, see the brilliant GM Watch 'Biotech Brigade' directory at: http://www.gmwatch.org/welcome.asp

#### International Service for the Acquisition of Agri-Biotech Applications (ISAAA)

The US-based International Service for the Acquisition of Agri-Biotech Applications was set up in 1991 and is funded by biotech companies to cajole developing countries to take up GM technology. Funders include Bayer CropScience, Monsanto, Syngenta, Pioneer Hi-Bred, and the BBSRC. ISAAA's multi-million dollar budget is matched by high-profile board members, past and present, including from Monsanto, Syngenta, and an adviser to the World Bank, but it has no representatives from farmer organisations in developing areas like Africa. The organisation's annual reports on the global uptake of GMOs are widely reported in the media. Yet, the accuracy of these figures and the purported farmer benefits are exaggerated,[27] sometimes on a truly epic scale – 20 times in the case of the area of GM cotton farming in South Africa.[28]

#### Biotechnology Industry Organization (BIO)

BIO is one of the main GM lobby groups in the US, and was created by the industry in 1993 to represent its interests in Washington. By 2002, it could boast the following anti-democratic statistics: a $30 million budget, 70 staff to do its dirty work, and 1,000 companies to do it for.[29] In the period from 1998 to 2002, BIO spent $14 million lobbying Congress, the FDA, and the White House.[30]

#### American Soybean Association (ASA) & National Corn Growers Association (NCGA)

The ASA gets almost 10 per cent of its budget from companies such as Monsanto, Pioneer Hi-Bred, and BASF.[31] The NCGA gets even more of its budget from Syngenta et al.[32] In 2001, the ASA spent $280,000 working with the NCGA and the Council for Biotechnology

Information to achieve 'a unified message about the benefits of transgenic crops'.[33]

### European Association for Bioindustries (EuropaBio):

This is a European lobby group established in 1996 and representing over 40 member companies worldwide and 13 national biotech associations.[34] According to Paul and Steinbrecher, in *Hungry Corporations*, it 'has proved to be a cunning and powerful lobbyist at the European Parliament and the European Commission'.[35] Its lobbyists 'regularly meet with Commissioners and organise dinner debates for MEPs, civil servants and academics'.[36] PR firm Burston-Marsteller rather worryingly stated that the organisation has an 'indispensable direct role in the policy-making process'.[37] According to the NGO CorporateWatch, it 'encourages the EU and national governments to develop policies that are supportive of biotechnology'.[38]

### AfricaBio

This group began in January 2000 in South Africa, and by December 2002 it had 90 members. The organisation states that its aim is 'to promote the enhancement of food, feed and fibre through the safe and responsible application of biotechnology'.[39] It lobbies for GM crops in Africa and beyond. But, according to GM Watch:

AfricaBio ... seeks to present itself not as a corporate lobby but as part of civil society. ... However, in one of its press releases it frankly stated that it was intended to 'provide one strong voice for lobbying the government on biotechnology and ensuring that unjustified trade barriers are not established which restrict its members'. It is clear from the list of AfricaBio's backers that in reality industry organisations dominate AfricaBio.[40]

An article in *Nature* provides some interesting insights about the group:

AfricaBio, along with agribiotech companies and other pro-biotech campaigners, is now fighting tooth and nail, often by somewhat controversial methods, to spread the word about GM crops ... [AfricaBio's] methods would be considered in some countries to be blatant media manipulation.[41]

### AusBiotech Ltd

This group was set up in 1995 as the Australian Biotechnology Association, changing its name in 2001 to AusBiotech Ltd. It has 685 members and is the chief biotech lobby group in Australia. 'Its mission', it states, 'is to facilitate the commercialisation of Australian bioscience in the international marketplace'. Its members include not only biotech corporations, but, interestingly, university and government research departments along with hospital departments.

### Agricultural Biotechnology Council (ABC)

The ABC was set up by the biotech industry in the UK, and is chaired by Stephen Smith of Syngenta. In February 2002, ABC's inaugural report concluded that if GM crops could be seen to be beneficial to birds, then the majority of people would support them. The ABC transferred its PR account to Lexington Communications, a nicely connected company whose director, Mike Craven, was chief of Tony Blair's press office. He also served with John Prescott, the deputy prime minister. Lexington now employs Bernard Marantelli, an ex Monsanto man, to look after a £250,000 campaign targeting 'regulators, legislators, retailers and consumer groups' to approve GM crops.[42]

### CropGen

An industry-funded lobby group operated by pro-GM UK academics. Its feedback to the May 2003 ActionAid report illustrates how much spin influences its statements. Not only did its press release deal with only one GM crop (cotton), but virtually every aspect of the document, according to GM Watch, one of the foremost anti-GM websites, was 'seriously open to question'. CropGen claimed average yield increases of 60 per cent with GM cotton, and a 70 per cent reduction in insecticide use 'resulting in environmental, social and economic benefits'.[43] This is a fascinating revelation, as it is a direct contradiction of all the independent reports on Bt cotton (see Resources, p. 236).

### Crop Protection Association (CPA)

Formerly known as the British Agrochemicals Association (BAA), this group is an agrochemical and biotech trade association, one of the organisations that make up SCIMAC, the industry body established in June 1998 to support the 'responsible and effective introduction of GM crops in the UK'. It's touching that the UK government is so trusting of industry, so much so that SCIMAC was allowed to be part involved in running the GM Farm Scale Evaluations (see pp. 121–3).[44]

### Sense About Science (SAS)

Sense About Science is one of a network of industry-financed groups (which also includes the next four organisations listed below) that has sprung up in recent years in the UK. The group is financed by pro-GM corporations like Unilever and GlaxoSmithKline,[45] and pro-GM groups like the Biotechnology and Biological Sciences Research Council (BBSRC) and the SIRC[46] (a pro-GM group that receives funding from the food and drinks industry[47]). SAS is one of the Royal Society's recent allies. Lord Dick Taverne, SAS's Chairman, took Monsanto's much criticised Much ado about a skylark road show to the House of Lords (see p. 137).[48] SAS director Tracey Brown is an ex-employee of PR company Regester Larkin, which includes among its clients biotech, pharmaceutical, and oil companies. An SAS group on scientific peer-review convenes at the Royal Society and includes Tracey Brown and the IoI's Tony Gilland.

### Institute of Ideas (IoI)

This is a libertarian institute with a curious past.[49] One of its funders is biotech corporation Novartis,[50] and it grew out of the left-wing *LM* magazine (which used to be called *Living Marxism*), published by Claire Fox, the media commentator. Both *LM* and the IoI have a track record of attacking environmentalists. Tony Gilland, of the IoI, considers the Farm Scale Evaluations were 'an unnecessary obstacle' to the establishment of 'beneficial and benign' GM.

### Science Media Centre (SMC)

Housed within the Royal Institution, the SMC claims to be 'an independent venture working to promote the voices, stories and views of the scientific community', but always champions GM and receives funding from the likes of DuPont, Pfizer, and Astra-Zeneca. Its head is Fiona Fox, formerly involved with *LM* magazine. Her sister is Claire Fox, of the IoI.

### Scientific Alliance (SA)

Set up in 1991 by the Director of the British Aggregates Association, Robert Durward, the group is corporate-funded and consistently anti-environmental and pro-GM and Big Business.[51] According to GM Watch, Robert Durward has described himself as 'a businessman who is totally fed up with all this environmental stuff', and who has written, 'Perhaps it is now time for Tony Blair to try the "fourth way": declare martial law and let the army sort out our schools, hospitals, and roads as well.'[52]

### International Policy Network (IPN)

Established in 2002, the IPN is an affiliation of global right-wing think tanks. The UK directors are Julian Morris (of the Institute of Economic Affairs – IEA) and Roger Bate. The group is also connected with the Sustainable Development Network; this is a coalition that includes Professor Prakash's AgBioWorld Foundation (see on), and which was formed in 2002, ready to push the pro-industry agenda at the upcoming World Summit on sustainable development.[53]

### Global network of pro-corporate activists

These lobby groups are just one part of a global network of pro-corporate activists, who are very vocal and ruthless in undermining their opposition. Their institutes and foundations sport a dizzying and conveniently confusing array of acronyms. Consequently, even independent journalists do little more than skirt round the edges of a subject, which is guaranteed to anaesthetise even their hardiest readers. Read on, then, at your own risk...

### Right-wing think tanks, research bodies, and PR companies

Big Tobacco finances quite a few of the members of this network. Philip Morris was actually on the board of Australia's Institute of Public Affairs (IPA), while the European Science and Environment Foundation (ESEF) was established largely with Big Tobacco money, despite claiming it is 'a non-aligned group of scientists' that does 'not accept outside funding.' ESEF is a front that assists tobacco companies in undermining research critical of industry, and attacks restrictions on tobacco, GM, and so on.[54]

The Competitive Enterprise Institute (CEI) is corporate-funded and pro-GM. Among its many sponsors are Philip Morris and Dow Chemicals.[55]

International Consumers for Civil Society (ICCS) was founded by Dennis Avery (and others) of the Hudson Institute (HI), and receives funding from the usual biotech interests, including Novartis, Cargill, DuPont, and Monsanto.[56] Dennis Avery is the author of many organic smears, mainly via a weekly column for newswire Bridge News, which disseminates his pieces to around 300–400 American newspapers.[57] Alex Avery, his son, works at the HI's Center for Global Food Studies.[58]

According to an article in the *Ecologist*:

In America, Avery's message has been picked up widely by other organisations, most prominently the American Council on Science and Health (ACSH) and The Advancement of Sound Science Coalition (TASSC). The ACSH is run by Dr Elizabeth Whelan, regarded as one of the top 50 'heroes' of the anti-environmental, pro-industry 'Wise Use' movement in the US. She too is on a 'crusade' against the "toxic terrorists" of the organic movement. Before they stopped revealing their sources, ACSH used to receive some 50% of their funding from some of the most 'toxic' corporations in the world, including Monsanto, Shell, Ciba-Geigy, Exxon, DuPont, and Union Carbide.[59]

Steven Milloy runs web pages attacking environmentalists and organic farming – www.junkscience.com, www.nomorescares.com, www.consumerdistorts.com, and StopLabelingLies.com. But Milloy's mission against 'junk science', from whence he got the tag the 'Junkman', was established with money from Philip Morris, in order to undermine research critical of industry.[60]

Graydon Forrer is an ex-Monsanto PR man and managing director of Life Sciences Strategies, a company whose speciality is 'communications programmes' for the life science and pharmaceutical industries.[61]

A US PR company, Berman & Co, was established with $600,000 from Philip Morris, the tobacco giant. 'Our offensive strategy', Berman admits, 'is to shoot the messenger. ... we've got to attack their [environmental activists'] credibility as spokespersons'.[62] According to George Monbiot in an article in the *Guardian*, the company operates a 'fake public interest site' entitled ActivistCash.com, which dissuades foundations from giving money to environmental campaigners. Berman also operates the Center for Consumer Freedom (CCF), 'which looks like a citizens' group', but is stridently anti-GM labelling. It also lobbies against restrictions, bans, and health warnings for drinks, tobacco, and fast food companies.[63] Michael Jacobson, the Director of the Center for Science in the Public Interest (CSPI), said: 'The so called Center for Consumer Freedom deceives the American people every day of the week by posing as a consumer group, when it's really a front group that does PR dirty work...'[64] The CCF promotes 'genetically improved food' and smears organic produce.[65]

According to GM Watch, Berman – in a 'genetically improved' version of the truth –

also paints biotech opponents as terrorists, asserting that 'anti-biotech extremists' are part of a 'growing wave of domestic terrorism' and that the people we need to worry about are not just al-Qa'ida but 'the middle-class kids down the street'[66] Berman's list of 'anti-biotech extremists' includes not just Greenpeace and Friends of the Earth ... but even organisations like Christian Aid,[67]

who spend millions helping to lift the poor out of poverty.

Nichols Dezenhall is a marketing firm with a 'citizen's initiative' website entitled StopEcoViolence, which admonishes activists. The company joined up with the corporate-funded and pro-GM Competitive Enterprise Institute (CEI), to back a conference on 'eco-extremism' for journalists and corporate officials.[68] According to an internal American Chemistry Council (ACC) memo, obtained by Environmental Working Group (EWG), this 'marketplace defence' firm, as it likes to call itself, 'hires former FBI and CIA agents, to conduct "selective intelligence gathering ... about the plans, motivations and allies of opposition activists"'.[69]

### UK anti-environmentalists

In the UK, there is a clique of extreme anti-environmentalists. After enormous defeats over GMOs, pro-GM lobbyists and a group of right-wing journalists and academics have bounced back to conduct an 'aggressive campaign of disinformation', according to GM Watch,

to smear GM critics and alternatives to agricultural biotechnology ... BBC coverage well illustrates what's been going on. Three BBC programmes in early 2000 gave prominence to extreme anti-organic views (*Costing the Earth*, *Counterblast*, and to a lesser extent the *Food and Drink* programme). ... A key contributor to each ... has been Julian Morris ... of the far right think tank, the Institute of Economic Affairs [IEA]. ... *Counterblast*, broadcast on BBC 2 on 31 January 2000, was presented by Roger Bate in his then role as the Director of the European Science and Environment Forum (ESEF).

The ESEF and IEA 'appear[s] to be extremely closely linked' – a fact not disclosed on the *Counterblast* programme.

What is more revealing is the way in which the BBC's science and technology unit, and senior academics like Prof Hillman or Prof Anthony Trewavas, another contributor to *Counterblast*, have apparently been happy to promote such views without serious critical scrutiny of the evidence on which they are based.

Bate and Morris's book, *Fearing Food. Risk, Health and Environment*,

contains a chapter on GM by Prof John Hillman, who has also engaged in highly dubious public criticism of organic agriculture. One of Hillman's co-authors is none other than Professor T. Michael Wilson, whose highly inaccurate public promotion of GM has also drawn criticism.[70]

### Anti-environmental authors

Another piece in the jigsaw puzzle of misinformation comprises the anti-environmental authors, for example Dennis Avery, with his *Saving the Planet with Pesticides and Plastic* and Roger Bate with *Global Warming or Hot Air?* And then there is Bjorn Lomborg, who, in January 2003, was found guilty of scientific dishonesty by the prestigious Danish Committee on Scientific Dishonesty, on account of his controversial anti-green critique, *The Sceptical Environmentalist*. The press failed to do its homework, and the book was favourably reviewed in much of the non-specialist media, especially the *Economist*, the *New York Times*, the *Sunday Times*, and the *Guardian*. At the 2002 Johannesburg Earth Summit, Lomborg was even given a slot on BBC 2 to expound his jaundiced theories. Jeff Harvey, a former editor of the prestigious journal *Nature*, said: 'Lomborg ... has based his conclusions on cherry-picking the studies he likes, and he has seriously undermined the public's understanding of important contemporary scientific issues.'[71]

### Prakash, AgBioView, and AgBioWorld

Professor C.S. Prakash's AgBioWorld foundation and website, whose *raison d'être* is the promotion of biotechnology, is probably the leading biotech website. It is an influential talking shop for GM scientists worldwide,[72] with new daily postings on the 'wonders' of GM crops, attacks on anti-GM science, and condemnations of environmentalists and scientists.[73] Interestingly, Tuskegee University, Alabama, where Prakash is Professor of Plant Molecular Genetics, receives multi-million dollar funding from the pro-GM US Agency for International Development (USAID).[74] According to GM Watch, much PR sleaze passes through Prakash's AgBioView list,[75] and

AgBioView's more extreme material accuses GM critics variously of fascism, communism, imperialism, nihilism, murder, corruption, terrorism, and even genocide; not to mention being worse than Hitler and on a par with the mass murderers who destroyed the World Trade Center![76]

Certainly, anti-GM activists know they've made the grade when they are slurred on either AgBioWorld or AgBioView. As we will see throughout the book, the unwelcome Prakash presence comes up with monotonous regularity. In the case of Monsanto's GM sweet potato in Kenya, he was caught out extolling the benefits of a crop that had already been shown to be a complete failure; questioned about this, he said he was unaware of these findings.[77]

## INTERNATIONAL BODIES AND GOVERNMENTS

### International bodies

A web of developed world 'clubs', notably the World Trade Organization (WTO) and Organization for Economic Co-operation and Development (OECD), 'exist to "open up" countries to "competitiveness", a current euphemism for plunder', as author, film-maker, and top journalist, John Pilger, puts it.[78] The three most important instruments of world economic power are the WTO, the World Bank, and the International Monetary Fund (IMF).

#### World Trade Organization (WTO)

The WTO, which was set up in 1995, is the main body regulating and policing world trade. Unelected and unaccountable, it has the authority – which must be very reassuring to Big Business – to override any national government.[79] Its business is highly secret, undemocratic, and cannot be challenged outside the organisation.[80] In 2001, it had 142 member countries.

The WTO is not a UN body and concerns itself only with trade liberalisation. This is highly damaging to poor countries, because it leads to cheap food imports and an increase in the prices of farm inputs, which is putting small farmers out of business all over the developing world. Under WTO rules, industrialised countries are allowed to continue agricultural protection, but developing countries, unable to afford such extravagancies historically, are not allowed to step up their support. In fact, the EU and the US have actually sharply increased support for agriculture from $182 billion in 1995 to $327 billion in 2000.[81] One rule for the rich, another for the poor.

According to a report by Friends of the Earth International (FoEI), August 2003:

Behind the rhetoric about 'rules-based trade,' 'liberalisation' ... the reality is that the WTO's trade and investment rules are consistently being shaped around the interests of transnational corporations, consolidating their global expansion and removing any remaining obstacles.[82]

John Pilger puts it another way:

The multinational corporations exercise their prerogative under the rules of the WTO to plunder poor nations of their resources, markets and labour, free from local interference and international accountability.[83]

In an agriculture context, Dan McGuire, of the American Corn Growers Association, said:

The WTO is being used by ... powerful, transnational agribusiness corporations to control the future of farm policy, trade policy, GMO policy, food production and the price you get for your commodities and their agenda is primarily to enhance their economic bottom line, not yours...[84]

George Monbiot observes: 'trade agreements have become the greatest threats to representative government on earth'.[85] He goes on to point out that WTO rules are 'among the foremost threats to equitable development, human rights, and the environment'. And that, rather frighteningly, 'World trade rules could, in principle, be used to dismantle the Montreal Protocol protecting the ozone layer, the Basel Convention on the shipment of hazardous wastes, and the Convention on International Trade in Endangered Species.'[86]

### World Bank and International Monetary Fund (IMF)

The World Bank – whose agenda is controlled by the countries that provide most of its funds (the US, the UK, Japan, Germany, and France)[87] – was set up in 1945 and is the world's largest and most controversial multilateral aid organisation.

Its total lending in 2000 was a very modest $15.3 billion spread over 180 member countries (which compares rather unfavourably to the much more impressive US military budget of $360 billion a year[88]). In return for such piffling loans, the World Bank forces free-market economies on to recipient countries, in the form of the tyrannical Structural Adjustment Programmes (SAPs). That means vast cutbacks on public services, privatisation of public companies, and very favourable terms for the multinationals, including cheap labour and 'tax holidays'. Vietnam is a classic example of a SAP in action. Within a decade, its exemplary education and primary healthcare systems were in tatters, unemployment had soared, poverty was ubiquitous, and pensions and social welfare for the most vulnerable were abolished.[89] A red light for democracy, green for corporate plunder – and all for a very reasonable price.

In Africa, SAPs applied by the US, together with the World Bank and IMF, have exacerbated famine. More than 600 million people on the continent suffer from malnutrition and a lack of basic healthcare and education because the money that should be spent on them is being used to repay the loans of the Western financial institutions.[90] And in the Philippines, one child dies every hour, while 50 per cent of the national budget is spent just on paying the interest on World Bank and IMF loans.[91] No region of the world is immune to this tyrannical logic. In 1995, severely indebted, low-income countries paid $1 billion more in debt and interest to the IMF than they received from it. According to UNICEF, 500,000 children die a year, because of the burden of unrepayable debt owed by their governments to the developed countries.[92]

As we have already seen with the Green Revolution, the IMF and World Bank have a dismal record on agriculture and related issues.

Between 1988 and 1995, the World Bank financed $250 million worth of pesticides, with two of the highly toxic 'Dirty Dozen' pesticides, Paraquat and DDT, appearing in these contracts. An outrageous $57 million of this money went to agrochemical manufacturers in G7 countries in just three years. And from 1993 to 1995, the World Bank, and other multinational development banks, in more acts of blatant corporate-welfare, channelled nearly $5 billion of aid money to that well-known neighbourhood of destitution – US big business. One of the chief beneficiaries was Cargill, the world's third biggest food company, whose annual sales are the size of the economies of countries like Pakistan or the Philippines.[93] The question is – who are the World Bank and IMF supposed to be helping?

The World Bank has for some time been quietly financing hundreds of millions of dollars worth of GM crop projects in developing countries. Dr Florence Wambugu's pitiful GM sweet potato flop show in Kenya, which cost $6 million,[94] is a classic example (see pp. 138–9). The World Bank's project partners include the usual suspects – Monsanto, Aventis, and Syngenta.[95]

Economist David Korten, a former adviser to the USAID, wrote: 'the World Bank and the IMF have been disastrous failures – imposing an enormous burden on the world's poor and seriously impeding their development'.[96] And 'They have arguably done more harm to more people than any other pair of non-military institutions in history.'[97] The Rev. Jesse Jackson put it rather more strongly, when addressing eleven African heads of state in Libreville in 1993: 'They no longer use bullets or rope. They use the World Bank and the IMF.'[98]

*Organization for Economic Co-operation and Development (OECD)*

The OECD has 30 member countries, and (like the WTO) one of its principal aims is 'trade liberalisation'. It has been working on biotechnology-related topics for some two decades, and is one of the leading international proponents of GM. Its Biotechnology Unit was headed from 1993 to 1998 by Mark Cantley, who promoted Planet Biotech's alarming view that GMOs are actually 'inherently safer and more precise' than non-GM seeds, vaccines, enzymes, pesticides, etc.

The 2000 and 2001 OECD conferences in Edinburgh and Bangkok were both chaired by GM proponents. Dr Pusztai was the only GM-sceptical scientist at the Edinburgh conference, and he was subjected to repeated and vociferous attack. He commented that 'it appeared to be more of a propaganda forum for airing the views and promoting the interests of the GM biotechnology industry'.[99]

*World Food Programme (WFP)*

The United Nations WFP is the world's largest international food aid organisation and is based in Rome. In 2000, it helped feed more than 83 million people in 83 countries. But nowadays – despite describing itself as 'the frontline UN organisation fighting to eradicate world hunger' – most of the WFP's work is in distributing food to keep people alive in emergencies, rather than helping people out of poverty.[100]

The WFP is very US-dominated, with the US providing more than 60 per cent of its aid.[101] In 1999, Catherine Bertini, a former official of the US Department of Agriculture (USDA), was its Executive Director.[102] In fact, the WFP went all the way with the US on trying to force GM food aid on southern Africa in 2002 (see p. 127).[103]

Food is power. We use it to change behaviour. Some may call that bribery. We do not apologise. – Catherine Bertini, Executive Director of the WFP, at the Beijing Woman's Conference[104]

*Food and Agriculture Organization (FAO)*

The Rome-based FAO is the largest specialised agency in the UN. It deals with agriculture, forestry, fisheries, and rural development, with a mandate to raise levels of nutrition and standards of living, to improve agricultural productivity, and to better the condition of rural populations.[105] But the FAO was a great proponent of the Green Revolution, which, as we have seen, created enormous problems for small farmers, people's health, and the environment in the developing countries.

In relation to GM, the FAO has been rather schizophrenic in its proclamations, given the manifold interests it represents. This is well illustrated by the 'GM will feed the starving' hype spouted by the GM lobby at every available opportunity (see pp. 49–50).[106] In 2000, a major report by the FAO stated that GM crops were not necessary to feed the world. Since then, it has come under enormous pressure from the US and the biotech industry, and not surprisingly, recent reports have come out in favour of GM, much to the disbelief and ire of aid agencies and environmental groups, who thought this tired old hype was long dead and buried.[107]

### World Health Organization (WHO)

As we will see later in the book, the WHO has been infiltrated by the food industry.[108] This is why the WHO so faithfully repeats the biotech industry line that all GM foods currently used have been assessed for safety and 'are not likely to present risks for human health'.[109]

### Consultative Group on International Agricultural Research (CGIAR)

CGIAR, the world's most important agricultural research system, is a worldwide network of research centres (IARCs) supported by an international donor group. Set up in the 1960s and 1970s, the broad objective of the centres was to increase the quantity and quality of food in the developing countries. In 1999, the CGIAR and its 16 centres (13 in the developing countries) had a budget of $349 million.

The chief criticisms of the IARCs are that in the 1960s the high-technology of the IARC-promoted Green Revolution helped larger farmers rather than poorer farmers;[110] and more recently, CGIAR has been accused of becoming a service centre for corporate interests, with an unhealthy enthusiasm for GMOs. It presently spends around $22 million a year on biotech R&D.[111] Despite severe criticism from its own non governmental organisation (NGO) advisory committee, it has welcomed Syngenta as a member of its governing board, thus having 'unabashedly adopted the corporate research agenda ... thereby ... accepting that it ceases to follow the original mandate of conducting agricultural research for "public good"', according to one NGO.[112]

According to Paul and Steinbrecher:

There is a great deal of evidence that the true beneficiaries of the CGIAR, both financially and in terms of germplasm, are the Northern industrial [developed] countries. Their returns on their investment in the CGIAR can be substantial...

Up to 1996, the US had gained around $17 billion in benefits from IARC-bred wheat alone. And large amounts of tropical seed samples end up in the hands of corporations like Pioneer Hi-Bred and Cargill, thus privatising the work of CGIAR.[113]

## US government

This book is littered with examples of the US administration leading the fight to force-feed the world GM foods. By way of a few examples of blatant pro-GM policy, the US government has:

- always opposed GM labelling laws both in the US and internationally, where, for example, it has put pressure on Mexico not to introduce mandatory GM labelling laws[114] and on the international GM labelling guidelines at Codex;[115]
- pressed for lax food safety testing on GM foods internationally;[116]
- resisted calls to reveal the location of secret biopharming test sites;[117]
- accelerated endeavours to transfer and distribute American agricultural technologies to Africa, particularly GM.[118]

### Food and Drug Administration (FDA)

The FDA is the agency with overall responsibility for the safety of GM foods, yet it appears to have rolled over and done whatever industry has asked of it, as we will see from the many examples in this book. In fact, Betty Martini, of the consumer group Mission Possible said: 'The Food and Drug Administration is so closely linked to the biotech industry now that it could be described as their Washington branch office.'[119]

### US Department of Agriculture (USDA)

Here's a taster of the long, long list of pro-GM actions by the USDA:

- The USDA attempted to change the legal definition of 'organic' to include GMOs, and to ban higher organic standards than its own, but it was forced to back down after getting 270,000 protest letters.[120]

- The USDA – along with USAID, and the US State Department – sponsored the $3 million 2003 Sacramento biotech conference, denounced by one Iowa farmer as 'a slick infomercial for Monsanto'.[121]
- The USDA launched an initiative in 2005 to help developers of GM fruits and vegetables 'work their way through federal regulatory requirements into the marketplace', since existing regulations were hindering their development.[122]
- The USDA continues to allow drugs and industrial chemicals to be engineered into food crops (biopharmaceuticals), with an advisory committee on the future of GM crops (including biopharmaceuticals) stacked with industry representatives.[123]

### US Agency for International Development (USAID)

According to the NGO GRAIN, USAID provides more than 60 per cent of WFP aid,

but it insists on either donating US foodstuffs or tying cash contributions to the purchase of US produce. Giving aid in kind alleviates the symptoms of famine but perpetuates the causes. This ... is part of a deliberate strategy to subsidise US agriculture and undermine its agricultural competitors. ... It is in the interests of the US ... to develop the South [the developing countries] only so much that it opens new markets and can purchase off the US. As Lawrence Goodwin, of the Africa Faith and Justice Network, observed: 'The US wants to see its corporations control life's most basic resources, including seeds, food crops and water.'[124]

Moreover, critics allege that USAID is used as an unofficial instrument of US foreign policy, forcing recipient countries to eat GM foods. In fact, USAID boldly states that one of its roles is to 'integrate GM into local food systems'.[125] In 2002, it launched a $100 million programme for bringing biotechnology to the developing countries – yet another biotech subsidy paid by the American taxpayer.[126] Ricardo Navarro, Salvadorian chairman of Friends of the Earth International, said: 'food aid [is] ... being used as a way to open up new markets for GM products'.[127]

Even the OECD and the World Bank have criticised USAID's self-serving agenda: 'Among the big donors, the US has the worst record for spending its aid budget on itself – 70% of its aid is spent on US goods and services.'[128] In fact, the USAID website itself boasts that 'the principal beneficiary of America's foreign assistance programs has always been the United States'. And that 'Foreign assistance programs have helped create major markets for agricultural goods, created new markets for American industrial exports and meant hundreds of thousands of jobs for Americans.'[129] USAID spends over $1 billion a year buying crops from US agricultural corporations and shipping them to the starving as aid.[130] This is a massive subsidy to US agribusiness, masquerading as altruism and paid for by the US taxpayer.

### UK government

The Blair Government is rabidly pro-GM, with an embarrassing habit of doing whatever the Americans ask.

Janet Bainbridge, former Chair of the Advisory Committee on Novel Foods and Processes, speaking like a true headmistress, sums up the government's attitude well. When asked whether people should be given the choice of eating GM or not, she replied that we should not because 'most people don't even know what a gene is... Sometimes you just have to tell people what's best for them.'[131]

As Dr Sue Mayer, of GeneWatch UK, observed: 'The government have never put consumers or the environment first. They have always put the industry and competitiveness first and that is just clear in all the policy documents and statements.'[132]

Michael Meacher suggested that the push to have GM foods on sale in the UK has been backed by 'senior people in government who are committed to the biotechnology industry'.[133] Meacher, who had been environment minister since the 1997 election, was sacked in June 2003 for being a lone voice daring to speak out against GM foods. And afterwards, as a backbencher, he became even more outspoken. Interestingly, the US administration consider him enough of a threat that the EPA has a file on him.[134]

Labour MP Alan Simpson adds weight to Meacher's words:

There is no doubt inside the corridors of power that the agenda is effectively being driven by [Lord] Sainsbury and the biotech corporations. It is being aided and abetted by the FSA [the Food Standards Agency], where Sir John Krebs [its then chief] is now unaffectionately known as 'GM Joe'.[135]

This is no doubt why the UK government has so often been the only country opposing anti-GM legislation in Europe. In June 1999, Britain blocked France and Greece's attempt to bring in a GM crop moratorium. It also lobbied and voted at the European Commission to stop Austria and Luxembourg from banning GM maize. And it used its 1998 EU presidency to scupper the labelling of GM derivatives.[136]

### Department of Environment, Food, and Rural Affairs (DEFRA)

DEFRA took over from the Ministry of Agriculture, Fisheries, and Food (MAFF) in 2001 after the debacle of Foot and Mouth (FMD). But, far from curtailing the activities of MAFF, many believe DEFRA was merely its reinvention.[137] MAFF always favoured the industrialisation of British farming, relentlessly pursuing chemical agriculture, mechanisation, and increasing farm size. Supporting GM was therefore a logical extension of policy for MAFF/DEFRA.

### Food Standards Agency (FSA)

The supposedly 'independent' FSA is anything but the champion of the consumer. The FSA has been accused of trying to weaken GM guidelines globally. It has also backed the US position on avoiding GM labelling and traceability of GM food and animal feed, a position condemned by the Consumers' Association, who 'remain bitterly disappointed at the anti-consumer stance the Food Standards Agency and UK government, as a whole, takes on this issue'.[138]

Under the leadership of Sir John Krebs (co-founder and non-executive Chairman of Oxford Risk Research and Analysis Ltd, whose work includes research and advice for the oil and pharmaceutical industries, and a wide range of other businesses[139]), the FSA was staunchly pro-GM. This was evidenced by its spending £120,000 of taxpayers' money[140] on its own parallel and highly misleading debate during the GM 'Public Debate', which was nothing short of an advertising bonanza for the biotech companies.[141] The FSA has also refused to get behind government policy to promote organic food and agriculture. In fact, Sir John brazenly admitted that his mission was to 'undermine' organic agriculture.[142] The Chief Executive of the Food Safety Authority of Ireland described Sir John's views as 'extreme'.[143]

Shortly after becoming its head, Sir John cooperated with a food industry-funded group, who set up a forum full of GM proponents, to establish enforceable guidelines on the media's reporting of contentious issues like GM foods – that is, to report them in a pro-GM light.

When it comes to GM research, the FSA has proven controversial. In one example, the FSA commissioned the Newcastle University feeding trial, which showed alarming problems with GM, and yet the

FSA, with its easy knack for misinformation, spun it into meaning GM foods were as safe as houses.[144]

### Advisory Committee on Releases to the Environment (ACRE)

ACRE is the UK government's statutory body that advises ministers on the risks of GMOs and whether consent should be granted for them. Yet, in 1998, eight out of thirteen of ACRE's members had close biotech links. Interestingly, by April 1999, it had given the green light to 160 applications for GM crop experimental planting and turned down precisely none.[145] ACRE's chairman is Professor Chris Pollock, who is Director of Research at the Institute of Grassland and Environmental Research (IGER), which has contractual and financial relationships with AstraZeneca and Aventis.[146]

### Advisory Committee on Novel Foods and Processes (ACNFP)

ACNFP is in charge of making sure that GM foods are safe for human consumption. It, too, is highly compromised by its links to the biotech industry.[147] In fact, Dr Williams, MP, said of ACNFP:

as I understand, all of the evidence [on the safety of GM foods] taken by the advisory committee comes from the commercial companies, all of that is unpublished. This is not democratic is it? ...the evidence comes, simply, in good faith, from the commercial companies?[148]

### Biotechnology and Biological Sciences Research Council (BBSRC)

The BBSRC was established in 1994 and is the public funding body for research and training in Britain's non-medical life sciences, with an annual budget of £220 million in 2002.[149] But with monotonous regularity, corporate links once again rear their ugly head. The Chairman of the BBSRC was, until 2002, the Chief Executive of Zeneca, and its main committees are stuffed with corporate representatives: Syngenta (sits on 3), AstraZeneca (on 2), GlaxoSmithKline (3), Pfizer (4), Unilever (2), not to mention Genetix plc, Lilly and Merck Sharp, Dohme, and Aventis.

No wonder biotechnology has been swallowing up the lion's share of the research funds – with £55 million given to just four GM research projects in 1999,[150] and a piffling £1.7 million given to organic farming.[151] No wonder that of the 255 food and farming research projects the BBSRC is funding, 26 are concerned with growing GM crops, and just one with organic production.[152]

The BBSRC has been accused of instituting what has been called 'a gagging order' via its code that prevents all publicly funded researchers from speaking out on concerns about GM foods. Disobeying leaves researchers open to being sued or dismissed – even retired staff, who are Fellows of BBSRC-funded institutes.[153]

### Department for International Development (DFID)

This department has been strongly criticised for its vigorous support of GM crops. It has been quietly funding a £13.4 million programme to create a new generation of GM animals, crops, and drugs in more than 24 countries throughout the developing world. Dr Sue Mayer, of GeneWatch UK, accused the DFID of 'deceiving' the public about the scale of the project. Included in the DFID schemes were projects linked to a controversial £65 million DFID aid programme in the Indian state of Andhra Pradesh, which will help push 20 million subsistence farmers off their land.[154] The DFID is one of the major promoters of GM in Africa, along with USAID and Monsanto-trained scientist Dr Florence Wambugu.[155]

### Other UK bodies

### National Farmers' Union (NFU)

In 1947, the NFU became the only representative industrial body in the country with which the government was bound to consult by law. It has often been described as 'the seat of the agricultural establishment', yet its ideology of food production at all costs has led to the extinction of a generation of skilled small farmers.

MAFF scarcely made a move without consulting the NFU.[156] In fact, both always sang from the same hymn sheet, with the NFU always favouring, and lobbying for, large farmers and agribusiness. The NFU, then, supports the government line that failing to adopt GM crops means losing out to the UK's North American rivals.

### John Innes Centre (JIC)

The John Innes Centre in Norwich, Europe's leading plant biotechnology institute, promotes itself as an impartial source of information and advice to government, the farming industry, educators, the media, and the general public, and yet it receives funding from just about every major biotech company. It houses the Sainsbury Laboratory and has (or has had) a research alliance with the following companies: DuPont, Zeneca, and Syngenta – although

Syngenta pulled out in 2002. Much bogus science and educational material issue from this 'impartial source' of information, which is cause for much concern, for in the JIC's own words the exposure it achieves in the local and national media is 'excellent'.[157]

The funding for the centre bears closer inspection. The JIC performs research for Zeneca and Lord Sainsbury, yet it benefits from taxpayers' money controlled by a director of Zeneca and Lord Sainsbury,[158] through the BBSRC, which gives it more than £10 million in taxpayers' money every year.[159]

### The Royal Society (RS)

The RS was formed in 1660 and is the world's oldest scientific organisation, and the most powerful one outside the US. It is the pillar of the scientific establishment and the scientific equivalent of the old boys' club. 'It remains a self-perpetuating elite,' says Moira Brown, Professor of Neurovirology at Glasgow University. 'Old, white, and male.'[160]

The once august Royal Society has gone from scientific impartiality to become one of the leading proponents of GM. Although primarily funded from the public purse, in recent years it has received millions, via its fundraising campaign, from major biotech interests.[161] Leading RS fellows also have commercial interests of their own.[162] According to Dr Tom Wakeford, an adviser to ActionAid on GM and sustainable agriculture in the Third World, this means that 'British citizens are paying taxes to fund an organisation that actively promotes the interests of multinational biotech corporations, under the guise of independent science.'[163]

The RS was central to the 1999 attack on Dr Pusztai over his ground-breaking research on rats fed with GM potatoes (see pp. 142–3). Its opening salvo was a review of Dr Pusztai's work (something it had never done before), followed by a critical letter published in the national press, on 22 February 1999, and signed by 19 RS Fellows.[164] In its February 2002 report on GM crops, the RS criticised his peer-reviewed, published research, using *non peer-reviewed, unpublished data*, it overlooked published studies, and it made no effort to get more research done.[165] As Pusztai so revealingly put it: 'Their remit was to screw me and they screwed me.'[166] Professor Lacey thought the Society's attack was 'absolutely grotesque'.[167] And the *Lancet* described the condemnation as 'a gesture of breathtaking impertinence'.[168]

In one of the most cynical opinion-rigging tactics of all time, they joined with the SIRC and the Royal Institution to produce guidelines telling journalists how to report science issues (read GM). The guidelines recommended a list of approved pro-GM 'experts' whom poor overstretched journalists can rely on for authoritative advice, including the likes of Tony Trewavas FRS, Professor of Applied Biochemistry at the University of Edinburgh[169] – the same Professor Trewavas who was named in the High Court in London in relation to a media libel case over GM (see pp. 145–6).[170]

A classic example of how the RS reports science was the Brooms Barn 'Much ado about a skylark' piece, which was published by the RS, then roundly slated by anyone who knew anything about skylarks (see p. 137).

The RS also stands accused of attempting to rig the science strand of the GM 'Public Debate'. Spokesman for GM-free Cymru, Dr Brian John, said:

For an organisation which supposedly exists to promote all that is best in British science, and which supposedly supports integrity and interaction, its direct involvement in pro-GM propaganda and in bad science is a disgrace. ... more to the point, it is turning the 'science strand' of the GM debate into a farce.[171]

Well, you've got to admit, they're an exceedingly active set of old, white men.

# 4

# Exposing the Wild Claims
# Made by the Biotech Lobby

Churning out an endless torrent of lies and misinformation must be a thoroughly exhausting business, but I guess you have no other choice when you've invested billions in a technology that doesn't work and that no one wants. No wonder, then, that the biotech lobby spends vast millions in the US alone on promoting GM.[1] The general tactic appears to be that if you tell enough lies, enough times, eventually it becomes received wisdom. Malcolm Hooper, Professor Emeritus of Medicinal Chemistry at the University of Sunderland, finds it all rather depressing: 'The thing that's dismayed me most is coming face-to-face with scientists who are not concerned about the truth.'[2]

So, here's a sort of Top Ten best lies used to justify GM.

## LIE NO. 1 GENETIC ENGINEERING IS THE SAME AS NATURAL PROCESSES

That genetic engineering is the same as natural change or traditional breeding is one of the fundamental lies that underpins the whole business of Genetic Trickery. I can only assume there were a lot of Class A drugs around when this idea came up. The transfer of genes from one species to another – for example putting fish genes in tomatoes, or using viral genes in crops, or creating biopharmaceutical crops – would never happen in nature. Scientists are playing God with the very stuff of life. Even the UK government admits this in its Genetically Modified Organisms Contained Use Regulations 2000 (CUR 2000), which define GM as 'the altering of the genetic material in an organism in a way that does not occur naturally by mating or natural recombination'.[3]

As Dr Eva Novotny, of Scientists for Global Responsibility, says:

There is no evidence that unlike species have ever crossed during the billions of years that life has existed on earth. If Nature tried this experiment, it must have failed. We must not be so arrogant as to assume that we are more clever than Nature, lest we precipitate an irreversible chain of biological evolution that ends in catastrophe.[4]

Even staff at the US Food and Drug Administration (FDA), the agency that gave GM the go-ahead, didn't agree with this claim. One FDA Compliance Officer observed that GM and conventional breeding were not the same and 'lead to different risks'.[5] And FDA scientists concurred, stating that in the case of GM 'natural biological barriers to breeding have been breached'.[6]

## LIE NO. 2 GENETIC ENGINEERING IS PRECISE AND SOPHISTICATED

Even the genetic scientists don't really believe this one. According to a Monsanto scientist, one of the two main methods of inserting a foreign gene into a crop plant cell is by firing a .22 calibre (gene) gun, loaded with small gold-tungsten pellets, which are coated with the genes of interest, at a piece of plant tissue. The DNA is 'physically wiped off those pellets and incorporated into the plant chromosome material', leaving 'a transformed cell, now containing our gene of interest'.[7] So, that's what they mean by precise and sophisticated, and there I was thinking minuscule tweezers and an electron microscope. The other method of gene insertion is by bacterial infection, which is prone to the same uncontrolled random insertion problem as the gene gun approach.

The impossibility of guiding a new gene leads to unpredictable effects. It has to be remembered that genes do not work in isolation – which is the kindergarten version of events the biotech lobby would like us to believe – but in highly complex relationships, which have evolved over millions of years. Any change to the DNA at any point will potentially affect its whole length in completely unknown and unpredictable ways. And the transposed gene may act differently in the new host, with the host's original genetic intelligence always being disrupted, to a lesser or greater degree. The new DNA frequently contains multiple copies, and these often induce gene-silencing; often copies are scrambled.[8]

According to Dr Michael Antoniou:

Disturbance of the gene function of the recipient plant can occur in a number of different ways. Firstly, the foreign gene can insert itself into the host's genes or their genetic switch-control elements, thereby disrupting their functions. Secondly, the gene transfer process in general is known to introduce, through as yet unknown mechanisms, hundreds or even thousands of additional mutational defects in the DNA, with potentially devastating consequences on global host gene function.[9]

Mutations are probably common, although the GM lobby like to keep this little embarrassment quiet. As Dr Arpad Pusztai observed:

In genetic engineering, a lot of GM plants never see the light of day, because for one reason or another they don't grow or they have an unpleasant colour like the GM salmon which turned green ... if GM is such a predictable, precise science, then you should be able to produce the same thing again and again. But you can't.[10]

More alarmingly, as the Natural Law Party Wessex points out:

obvious adverse mutations can be easily 'weeded' out. But what about the abnormalities that cannot be seen by eye? Very often many of these will go undetected simply because little or no molecular research has been done as to their existence, nature and significance.[11]

### LIE NO. 3 GM CROPS HAVE BEEN ADEQUATELY TESTED

As we saw in Chapter 2, not only are GM crops inadequately tested, there has been only one minor safety test of GM food on humans. The industry, on the other hand, cites hundreds of safety tests, the vast majority of which are a complete red herring.

### LIE NO. 4 GM CROPS ARE ADEQUATELY REGULATED

Clearly, Chapter 2 shows also that GM crops are not well regulated. This is particularly so in countries like America, where GM crops have been commercialised. However, even in the EU, which has extensive GM regulation, there are still enough loopholes to allow significant amounts of GM material into our food.

### LIE NO. 5 GM CROPS ARE SAFE

The argument that GM crops are safe, because 300 million Americans have eaten GM food for nine years with no ill effects, is as scientifically absurd and intellectually lame as suggesting that we're safe in the hands of George W. Bush because he hasn't plunged us into World War Three – well, not yet anyway.

What we do know is that in the US, food-derived illnesses have doubled over seven years, coinciding with the introduction of GM food.[12] Obesity has rocketed, with the incidence of lymphatic cancers and many other illnesses rising.[13] In the UK, there was a 50 per cent increase in soy allergies reported when GM soy imports began.[14] And in Russia, the number of people with marked symptoms of allergy tripled in three years, possibly because GM food and organisms are widely traded across Russia.[15]

We don't know whether or not all this is due to GM foods, because nobody is looking. As Dr Vyvyan Howard, who runs the Infant Toxicopathology department at Liverpool University, said:

there is no baseline data, no exposure data, no human feeding trials, so it's an uncontrolled experiment! If GM foods are causing changes to common conditions (allergy, cancer, auto-immune disease) there is no way we could know.[16]

Ben Miflin, former Director of the Institute of Arable Crops at Rothamsted, near London, and a proponent of GM crops, argues that, under current monitoring conditions, any unanticipated health impact of such foods would need to be a 'monumental disaster' to be detectable.[17]

Donella H. Meadows, Adjunct Professor of Environmental Studies at Dartmouth College, observes:

Next time you hear a scientist asserting that gene splicing [GE] is safe, remind yourself that there is no scientific evidence for that statement. We are profoundly ignorant about what we are doing to the code that generates all life. And unfortunately some scientists, including those entrusted with public safety, are willing to lie.[18]

Geneticist David Suzuki puts it even more strongly: 'Any politician or scientist who tells you these products are safe is either very stupid or lying. The experiments have simply not been done.' The Royal Society of Canada said, in January 2001, that it was 'scientifically unjustifiable' to assume that GM foods are safe.[19] Japanese scientist, T. Inose, observed that 'side-effects and accidents are inevitable, and scientists have assessed the risks from GM-foods and crops as unlimited'.[20]

And Dr Suzanne Wuerthele, a scientist specialising in risk assessment, said:

It took us 60 years to realize that DDT might have oestrogenic activities and affect humans, but we are now being asked to believe that everything is OK with GE foods because we haven't seen any dead bodies yet.[21]

## LIE NO. 6 GM CROPS WILL HELP FEED THE HUNGRY MILLIONS

This is a world-class whopper of a lie, which keeps running and running. And it's not just the biotech companies, but President Bush himself using this to force-feed the world America's GM leftovers, like a champion foie-gras stuffer. To the layman, this snappy phrase sounds plausible, even laudable. Sadly, the truth is much more complex and much harder to get across in a sound-bite age.

The truth is, there is more than enough food produced to feed everyone in the world – without GM crops.[22] People go hungry because of poverty, because they cannot buy from the plenty around them, not because of a lack of GM seeds. 'What poor people really need is access to land, water, better roads to get their crops to market, education and credit schemes,' says Matthew Lockwood, ActionAid's Head of Policy. 'GM does not provide a magic bullet solution to world hunger.'[23]

Representatives of developing countries abhor this cynical emotional blackmail. Delegates from 20 African countries to the UN's Food and Agriculture Organization in 1998 stated:

We strongly object that the image of the poor and hungry from our countries is being used by giant multinational corporations to push a technology that is neither safe, environmentally friendly nor economically beneficial to us. ... On the contrary, we think it will destroy the diversity, the local knowledge and the sustainable agricultural systems that our farmers have developed for millennia and that it will thus undermine our capacity to feed ourselves.[24]

ActionAid states the agenda of the multinationals more strongly:

It is not the interests of poor farmers but the profits of the agrochemical industry that have been the driving force behind the emergence of GM agriculture.[25]

If the biotech corporations care so much about the poor, why is only 1 per cent of GM research aimed at crops used by poor farmers?[26] Furthermore, GM technology might offer developed countries the opportunity to grow crops that can substitute those currently grown in the developing countries, such as plant oils, sugar, vanilla, cocoa butter, and flavourings.

And why are the biotech corporations patenting seeds? This will mean the poorest – that is the 1.4 billion subsistence farmers[27] – will not be able to save their seeds from year to year. This practice would become illegal and farmers would become totally dependent on the corporations for their seeds. In other words, they'll have to pay top dollar for them, rather than getting them for free. 'Increased hunger [in the developing countries] will be the impact of patenting seeds,' says Roger Dubois, president of Peace and Development, a Catholic agency active in the developing countries.[28]

Besides, the so-called Gene Revolution, like the Green Revolution before it, will not reach the majority of the poor, the subsistence farmers, because they are unable to afford the expensive GM seeds and inputs. It will just further intensify the process of agricultural industrialisation – which is all the biotech industry is really interested in. Likewise, it will also cause many of the problems that the Green Revolution did.

And if all this sounds like a commercial idea searching for a philanthropic cause, GM yields are usually no more than non-GM, and often lower – whereas sustainable agriculture provides enormous yield increases in the developing countries and is far more relevant and accessible to the poor. Thus, money invested in GM technologies is scarce money and time not invested in sustainable agriculture.

GM will not feed the poor; rather the poor will feed GM shareholders.

### LIE NO. 7 GM CROPS WILL INCREASE FARMERS' INCOMES

Report after report (see pp. 234–6) shows that this is just another fairytale from the GM lobby. For example:

- A 2003 study from Professor Caroline Saunders at Lincoln University, New Zealand, says GM food releases have not benefited producers anywhere in the world.[29]
- The prestigious June 2002 USDA report on GM crops, grown in the US, states that GM Bt[*] insecticide corn had a negative economic impact on farms, while the use of Herbicide Tolerant (HT or Roundup Ready/RR) soybeans had no statistically significant effect on net returns. It concludes that 'Perhaps the biggest issue raised by these results is how to explain the rapid adoption of GE crops when farm financial impacts appear to be mixed or even negative.'[30]
- The Soil Association's 2002 'Seeds of Doubt' report, created with feedback from farmers, and data from six years of commercial

---

[*] GM insecticide resistant crops have received a gene from the soil bacterium *Bacillus thuringiensis*, which causes the crop to create its own pesticide – *Bacillus thuringiensis* toxin or Bt.

growing in North America, shows that GM soy and maize crops deliver less income to farmers (on average) than non-GM crops.[31]

- Iowa State University agricultural economist Michael Duffy concludes from his 2002 study that the primary beneficiaries of GM herbicide-resistant (RR) soy and Bt corn are not farmers.[32]

Some of the reasons for these reduced incomes are: lower market prices for GM crops (13 per cent more for non-GM soy), lower yields, higher costs for GM seeds (about 25–40 per cent more their non-GM counterparts),[33] and contamination problems.

Jane Doe, an Iowa soybean farmer and a member of Iowa Farmers, said:

GMOs have been a legal, environmental and financial disaster for American farmers. This report ['Seeds of Doubt'] is overwhelming proof that farmers have everything to lose and little to gain by growing GMO crops.[34]

John Kinsman, vice-president of the National Family Farm Coalition and dairy farmer in Wisconsin, said:

GMOs do not provide a quick fix solution to the economic problems of US farmers. As time goes on the technology is proving to be more of a hindrance than a help.[35]

Meanwhile, a classic Monsanto magazine ad from 2002 proclaimed:

There's profit in your fields. Unleash it with Asgrow Roundup Ready soybeans... With Asgrow soybeans, profitability runs wild.[36]

## LIE NO. 8 GM CROPS WILL BE HIGHER-YIELDING

According to the USDA report, GM crops *do not* increase yield potential and may actually reduce yields. Many other independent reports from around the world back this up:

- In the US, peer-reviewed research at the University of Nebraska has confirmed the poor yield performance of RR soy, the world's biggest GM crop, as between 6 and 11 per cent lower than non-GM cultivars.[37]
- In Argentina, GM soy yields are 5–6 per cent less.[38]

- In the UK, reports of crop trials from the UK NIAB showed yields from GM winter oilseed rape and sugar beet 5–8 per cent less than high-yielding conventional varieties.[39]
- In North America, the Soil Association report states that RR soy and RR oilseed rape produced lower yields than non-GM varieties on average, and although Bt maize produced a small yield increase overall, it was not enough over the whole period to cover the higher production costs.

In 1998, Bill Christison, who farms over 2,000 acres in Missouri, and is President of the US National Family Farm Coalition, said this about RR soy: 'The promise was that you could use less chemicals and produce a greater yield. But let me tell you none of this is true.'[40]

Furthermore, a Mississippi state court ruled in September 2001 that Monsanto's RR soy seeds were responsible for the reduced yields obtained by Mississippi farmer Newell Simrall. They awarded him $165,742 in damages.[41]

Again, all of this is in rather wonderful contrast to the advertising hype from Monsanto et al. For example, an advert for Monsanto's Asgrow soy in 2002 stated: 'Asgrow varieties return more and yield higher because they're driven by progress.'[42] And Monsanto literature, from 1998, states that the company had achieved an average 5 per cent yield increase with its RR soy.[43]

### LIE NO. 9 GM CROPS WILL REDUCE PESTICIDE USAGE

Utter fairytale.

Compare Monsanto's claim – consistent to a fault in their misinformation – that 'herbicide use was, on average, lower in Roundup Ready soyabean fields that in other US soyabean fields' and that a 22 per cent reduction could be expected,[44] to report after report showing that GM crops cannot be depended upon to reduce pesticide use, and often increase it. For example:

- The USDA report states that GM HT crops (those which are tolerant to certain herbicides – this includes RR crops) have produced no reduction in herbicide use, and that for RR soy herbicide use has increased.[45]
- The Soil Association report shows that RR soy, RR maize, Bt maize, and HT rape have mostly resulted in an increase in agrochemical use.[46]

- And most conclusive of all is the November 2003 report from agricultural economics consultant Dr Charles Benbrook. This is the first comprehensive study of the impacts of all major GM crops on pesticide use in the US over its first eight years of commercial use. It shows that the planting of 550 million acres of GM corn, soybeans, and cotton in the US since 1996 has increased pesticide use by about 50 million lbs (c. 22,700 tonnes).[47]

One of the main reasons for the increase in pesticide usage is because GM crops confer their herbicide resistance, through contamination, to weed species within just a few years. This means that glyphosate (the active ingredient in the herbicide Roundup) often has to be used at much higher doses as a result of GM crops – 6 to 13 times more in the case of resistant horseweed in Mississippi.[48]

According to the Soil Association report: GM crops are also resulting in 'the return to older, more toxic chemicals [sometimes in cocktails], such as Paraquat and 2,4-D [and atrazine], the very chemicals that GM crops were supposed to render obsolete'.[49] This happens as weed resistance to glyphosate and glufosinate increases.[50]

But Monsanto, who never miss a trick, has decided to turn this industry-made disaster into a new marketing opportunity. In June 2001, the company took out a US patent on mixtures of chemicals, including – incredibly – the tank mixtures that farmers make from mixing bought chemicals into cocktails on farm.[51] This is about as logical as food manufacturers patenting home-made cocktails like water, sugar, milk, and a teabag. This 'will make it even more difficult and expensive for farmers to control [industry-created] HT volunteers,' observes the Soil Association report.[52] Volunteers are crop seeds, which drop to the ground, ready to germinate into a future crop. If the volunteer and the new crop are different crops, the volunteers then become weeds. And, if the volunteers are a GM HT variety, for example Roundup Ready, then Roundup herbicide used to kill weeds in this crop will not kill them.

Patently, this reduction-of-pesticide hype is nonsense. Why would an agrochemical corporation want to do that? The fact is that Monsanto's glyphosate (Roundup) patent ran out in 2000, which would have meant a massive loss of revenue, since half the company's agricultural revenue (a colossal $2.8 billion a year) came from the chemical.[53] GM Roundup Ready crops were the centrepiece of Monsanto's continued glyphosate sales, with farmers contractually obliged to buy the company's own brand of glyphosate, that is Roundup.[54] While telling everyone that Roundup Ready crops would reduce herbicide use, the company was quietly cranking up glyphosate production ready for their release.[55]

## LIE NO. 10 GM CROPS MUST BE SUCCESSFUL AS THEY ARE WIDELY GROWN

The biotech lobby omit to tell us the many less flattering reasons why GM crops are being grown. The Soil Association's 'Seeds of Doubt' report, however, does, with the following list of reasons that are luring many farmers into growing GM crops, who otherwise wouldn't:

- A lack of independent facts.
- A lack of awareness of the market rejection abroad.
- A lack of awareness of the agricultural problems. The gagging orders used by the biotech companies, after patent infringement allegations, have hidden the scale of these problems.
- The fact that some farmers weren't even told they were buying GM, initially, merely a new hybrid.
- The herbicide price war – which roughly coincided with the commercialisation of GM crops, and allegedly involved selling subsidised agrochemicals to growers – has balanced out the higher herbicide use costs.
- The substantial extra subsidies from the US government have masked the economic problems of GM crops.
- The 'lock-in' effect has made it hard for those farmers growing GM crops to stop. Factors include a lack of decent non-GM varieties, risks of crop contamination, the inability to access valuable GM-free markets, and the possibility of what could be very costly patent infringement accusations by Monsanto (see pp. 87–9).[56]
- One of the industry's greatest marketing assets, in North America, was farmer ignorance, exacerbated by a very bad economic situation. According to Shannon Story, women's president of the Canadian NFU: 'the increase in acreage is the result, more than anything, of a lot of salesmanship'.[57]

Lyle Wright, a Canadian NFU member, puts it this way: 'The promises come first and only later come the realities of contamination and genetic pollution, higher seed costs, market loss and superweeds.'

Stewart Wells, President of Canada's NFU said, 'UK farmers should not be fooled.'[58]

## LIE NO. 11 GM CROPS WILL HELP THE ENVIRONMENT

Professor T. Michael Wilson, formerly of the Scottish Crops Research Institute (and consultant for Lord Sainsbury's biotech investment firm Diatech),[59] said in 1997:

To feed 10.8 billion people by 2050 will require us to convert 15 million square miles of virgin forest, wilderness, and marginal land into agro-chemical dependent arable land. GM crops hold the most important key to solve future [environmental] problems in feeding an extra 5 billion mouths over the next 50 years.[60]

This contrasts starkly with the real-life experience of Argentina, where: '[much] pampas ... has disappeared, as have hundreds of thousands of hectares of forest',[61] leading to 'floods without precedence' as these habitats are replaced with GM soy.[62] Furthermore, Argentina is now being doused with ten times more glyphosate herbicide than before GMOs were introduced.[63]

Dr Jules Pretty, Professor of Environment and Society, University of Essex, is not impressed by Michael Wilson's assertion either:

To anyone who comes from developing countries or has worked in them, this is simply nonsense. Time and time again, we found that farmers who shifted to well-managed sustainable organic systems succeeded in both saving money and increasing yields. For instance, 100,000 small coffee farmers in Mexico who adopted fully organic methods have increased yields by half.[64]

In the developed world, GM is about as useful to the environment as a good old-fashioned nuclear accident. Graham Wynne, Chief Executive of the UK's Royal Society for the Protection of Birds, observes:

The ability to clear fields of all weeds using powerful herbicides which can be sprayed onto GE herbicide-resistant crops will result in farmlands devoid of wildlife and will spell disaster for ... already declining [species of] birds and plants.[65]

Research published by English Nature (the UK government's environmental adviser) in 2003 indicated that GM rape and sugar beet could make birds such as the skylark extinct within 20 years.[66]

Even the claim by the biotech brigade that GM crops will help the environment by increasing no-till agriculture is conveniently flawed, since the USDA report states that using HT seed did not significantly increase no-till agriculture.[67] This practice depends on chemical, rather than mechanical, weed control and is used to help control soil erosion in the American Midwest. One of its main principles is sowing after minimal or no cultivation of the soil, as opposed to after a deep cultivation method like ploughing. But since when has a practice dependent on herbicide use been a benefit to the environment?

### LIE NO. 12 GM CROPS WILL HELP FARMERS USE MARGINAL LAND

That GM crops will help farmers use marginal land, and so protect forests and precious ecosystems, is another bit of biotech window dressing. Virtually 100 per cent of all GM crops are grown either for herbicide tolerance (70 per cent) – so the biotech companies can sell even more weed killers – or insect resistance characteristics (30 per cent).[68] Neither helps farmers use marginal land. In fact, many people believe engineering characteristics like drought-tolerance would be extremely difficult because of the number of genes involved.[69]

But plenty of non-GM solutions exist – as we will see later in the book – for example, hundreds of non-GM rice varieties tolerant of flooding, drought, and salinity.

There are also many ways in which organic or sustainable farming revitalises marginal soils. As Professor Jules Pretty observes:

Sustainable agriculture starts with the soil by seeking to reduce soil erosion and to make improvements to soil physical structure, organic matter content, water-holding capacity and nutrient balances.[70]

In developing countries, organic farmers use trees, shrubs, and leguminous plants to stabilise and feed soil, dung and compost to provide nutrients, and terracing or check dams to prevent erosion and conserve groundwater. The mucuna bean, in Honduras – which produces massive amounts of weed-suppressing growth – has created rich, friable soils on steep hillsides with degraded soils in only two to three years. The bean is a legume, and so fixes nitrogen from the air, thus also improving soil fertility. Maize is planted in the mulch after the beans are cut down. In time, both crops are grown together. Grain yields soon double or even treble.[71]

Furthermore, the whole issue of using marginal land is much more complicated than the intellectually challenged slogans of the biotech industry imply. The need to cultivate marginal lands often only arises from the displacement of farmers from the best land by urbanisation and the industrial growing of cash crops, often by multinationals. In Brazil, more land is owned by multinationals than all the subsistence farmers put together.[72]

## LIE NO. 13 GM GOLDEN RICE WILL SOLVE VITAMIN A DEFICIENCY

Golden rice (or pro-Vitamin A enhanced rice) was developed in Europe and made its debut in Asia in January 2001, when Manila's International Rice Research Institute (IRRI), Syngenta, and the Rockefeller Foundation announced the arrival of the first research samples. The biotech lobby has continually proclaimed that GM Golden rice, with its enhanced levels of beta-carotene, which are converted to Vitamin A in the body, will help solve Vitamin A Deficiency (VAD). This is a largely preventable problem occurring mainly in poor countries, which affects 124 million children, killing one million of them a year and causing irreversible blindness in a further 500,000.[73]

The biotech hype really is ludicrous, though. There is so little Vitamin A in this gimmicky product that an adult would have to eat so much to get their daily Recommended Daily Allowance of Vitamin A (9 kg of cooked rice a day) that it would probably kill them.[74] On a more realistic note, fruit, vegetables, and greens have hundreds of times more Vitamin A than Golden rice and these can be grown in every backyard. Coriander, drumstick, and curry leaves contain almost 14,000 microgram/100 gm of beta-carotene (a precursor to Vitamin A) in comparison with Golden rice, which has a minuscule 30 microgram/100 gm. Even the farmer-bred non-GM 'Red rice' of Uttaranchal contains more beta-carotene.[75] Furthermore, the bio-availability of the Vitamin A would be low, as its absorption is dependent on other factors not addressed by Golden rice.[76]

'This whole project is actually based on what can only be characterized as intentional deception,' said Benedikt Haerlin, Greenpeace's former international GM Coordinator.[77] Even the President of the Rockefeller Foundation, one of its funders, said: 'the public-relations uses of Golden rice have gone too far'. And 'We do not consider Golden rice the solution to the Vitamin A Deficiency problem.'[78] Greenpeace's Michael Khoo observes that Golden rice 'isn't about solving childhood blindness, it's about solving biotech's public relations problem'.[79]

Like feeding the poor, and using marginal soils, the problem of VAD is not simply a case of inventing a wonder food (which this certainly is not) – it is much more complicated. Micronutrient deficiencies are symptoms of poverty, poor hygiene, environmental degradation, and social inequality.[80]

Besides, it's all a colossal waste of money. Much cheaper options exist. For example, a high potency tablet can be given twice a year to children at risk of blindness. This costs $0.05 per tablet – which means that it would only cost $25,000 to save the eyesight of 500,000 children every year. By contrast, Golden rice has cost over $100 million to date, and it is not even ready yet – another great value-for-money product from the biotech industry.

## LIE NO. 14 THE GM 'PROTATO' WILL SOLVE PROTEIN DEFICIENCY

The GM protein-enhanced potato, the 'Protato', is being sold on the basis of improving Indian school children's nutritional intake. Yet, the GM potato contains a dismal 2.5 per cent protein, while the non-GM crisp potato has 6 per cent, the amaranth grain has about 16 per cent,[81] and pulses on an average contain 20–24 per cent. Even wheat contains 8–9 per cent.[82] Fervent GM supporter Professor C. Kameswara Rao said: 'I do not understand how this dismal product could generate so much euphoria.'[83]

# 5

# The Risks and Dangers of GMOs

For a new 'science', touted as the solution to many of the world's problems, GM is turning out to be the Mr Bean of modern technology, as we are about to see from the long, long list of problems that have followed in its wake.

## PROBLEMS GROWING GM CROPS

### Contamination

According to the Soil Association 2002 'Seeds of Doubt' report:

The single greatest problem that GMOs have caused in North America is the widespread contamination of the agricultural and food sector. It has happened within a few years of the commercialisation of GM crops and has occurred at all levels of the food chain, from seed and crop production to food manufacturing.[1]

The biotech industry has created the problem, and reaped vast profits from GM, yet the entire food industry is having to bear the costs.[2]

#### Biological mechanisms of contamination

*Horizontal Gene Transfer (HGT):* HGT is the transfer of genetic material 'between sexually unrelated and incompatible organisms and is known to have occurred across the boundaries of species, genera and even kingdoms'. It occurs directly from one individual to another, by processes similar to infection, as opposed to the normal process of vertical gene transfer that occurs in reproduction, from parents to offspring. Notable HGT hotspots are the gut and the plant root zone (rhizosphere), which are rich in nutrients; and HGT usually takes place with micro-organisms, especially bacteria.

There is evidence that GM plants may present a higher risk of HGT than non-GM plants, where HGT is uncommon.[3] The problem with GM is that any HGT that takes place will spread potentially dangerous genes into organisms that would not normally have them, with potential health and agronomic problems. Examples of this are in the exchange of: Antibiotic Resistant Marker (ARM) genes between a

GM plant and gut bacteria (creating new antibiotic resistant bacteria); herbicide resistance between GM plants and weeds (creating new herbicide resistant weeds); viral material between a GM plant and an infecting virus (creating more virulent strains or species of viruses).

*Cross pollination:* Cross pollination works within related organisms, through which genetically engineered information can easily spread. Pro-industry sources have always claimed that contamination distances were negligible, with buffer zones for rape set at a derisory 200 m in the UK crop trials. Judith Jordan (later Rylott) of AgrEvo (now Bayer) gave evidence under oath that the chances of cross pollination beyond 50 m were as likely as getting pregnant from a lavatory seat.[4] Well, you have been warned. But oilseed rape pollen has been found to travel 26 km,[5] maize pollen 5 km,[6] and GM grass pollen 21 km.[7]

The biotech industry has also claimed that hybrids from GM contamination die out quickly, but research shows that this is not necessarily the case.[8] It is also interesting to note that, compared to conventional plants with the same gene, scientists found that GM Herbicide Tolerant (HT) mustard plants were 20 times more likely to breed with related species.[9]

### Seed contamination

During the breeding and production of seed, GM contamination occurs by gene transfer (that is, by HGT or cross pollination), accidental seed mixing, or the use of soiled machinery. As a result, the US organic certifier believes that North American farmers can no longer source GM-free maize, rape, and soy seed, because GM contamination is so pervasive. The Canadian Seed Trade Association says that all non-GM varieties (where there is a GM variety) are contaminated on average with 1 per cent GM seed.[10]

### Crop contamination

Crop contamination occurs by the same means as seed contamination, with some additions – such as seed blown off passing trucks or any number of natural methods, like wind, rain, birds, animals.

In the Canadian province of Saskatchewan, the province with most organic farming, the majority of organic growers have abandoned growing canola because of the level of GM contamination. Their complaint has been taken to court.[11] Traces of GM grains, especially soybeans and corn, are repeatedly creeping into US wheat supplies.

Similar problems are surfacing in Australia.[12] And in 2004, massive GM contamination of non-GM papayas was found on Hawaii, with 50 per cent contamination of tested seeds from across the Big Island, of which 80 per cent were from organic farms.[13]

Crop contamination problems are normal in GM-growing nations, but more worryingly, they are also occurring in non-GM nations. In Greece, despite the official ban on cultivating GM crops, independent testing found high quantities of GM cotton in 2003's harvest.[14]

### Superweeds and Roundup resistant weeds

Where horizontal gene transfer or cross pollination takes place between HT crops and weeds (this includes Roundup Ready (RR) varieties), herbicide resistant weeds are created. These are known as superweeds. As a result, farmers are having to use older and stronger herbicides, often in cocktails (that is more), to eradicate these weeds. As Martin Entz, Professor of Agronomy at the University of Manitoba, puts it:

GM canola [now a superweed] has, in fact, spread much more rapidly than we thought it would. It's absolutely impossible to control... It's been a great wake-up call about the side effects of these GM technologies.[15]

Martin Phillipson, Canadian Professor of Law, is not impressed either:

Does Monsanto have any liability for this technology? Farmers in this province are spending tens of thousands of dollars trying to get rid of this canola that they didn't plant. They have to use more and more powerful pesticides to get rid of this technology, and Monsanto seems to have no liability.[16]

Worse still is multiple resistance – or Gene Stacking – where weeds or volunteers (a previous year's crop seed, which drops to the ground and germinates into a future crop) are contaminated with herbicide resistant genes for several different herbicides. Three or more is not uncommon in GM rape. In Canada and the US, multi-herbicide resistance occurred within three to four years.[17]

The GM monster hasn't taken long to show its true colours. And yet, according to the Soil Association, the biotech industry has repeatedly dismissed such concerns, despite all the evidence.[18]

Roundup resistant weeds are plaguing GM soy and cotton fields in the US. For example, Roundup resistant mare's tail, a prolific seed producer, whose seeds are easily blown by the wind, has been found in Delaware, Tennessee, Kentucky, Indiana, and Ohio. It now affects around 200,000 acres of soybeans in west Tennessee alone and 36 per cent of all cotton in the state.[19] In fact, 'the list of glyphosate [the active ingredient in Roundup] resistant weeds is getting longer and longer', according to an ad on the Delta Farm Press website. 'Waterhemp, Rigid Rye Grass, Mare's Tail, Italian Rye Grass, Velvetleaf' are also listed, as are signs of resistance emerging in ragweed.[20] So serious is the problem that an independent study says Roundup resistance is set to wipe around 17 per cent off US farmland rentals. What's more, 46 per cent of the farm managers surveyed said weed resistance to Roundup is now their top weed-resistance concern.[21] All of this is thanks to the great pioneering work of Monsanto.

Unfortunately, though, the problem is not just restricted to Roundup resistant weeds – the same is happening with glufosinate resistant varieties too.[22] And weed resistance is not just restricted to North America; in Western Australia, herbicide resistance in weeds is having a big impact on productivity too.[23]

### Threatening non-GM and organic agriculture

All this shows that 'coexistence', the cosy industry term for growing GM and non-GM crops nearby, is another genetically modified joke. GM genes will pollute non-GM and organic crops and wild relatives for miles around. A European Commission report concluded that if only 10 per cent of a country or region was planted with GM crops, the resulting contamination would probably mean that 'organic farms will lose their organic status'.[24] A recent study in the US showed that more than two-thirds of conventional crops are now GM-contaminated – dooming non-GM agriculture and threatening people's health.[25]

In Britain, production costs for conventional and organic oilseed rape and maize would increase by 41 per cent if GM was commercialised, in response to problems of GM contamination.[26] And in Australia, Bob Phelps of the Genetics Network estimates that 'to segregate and certify GE-free grain will cost farmers 10 to 15 per cent of their crop's value... If they don't, they will lose markets in the Middle East, Europe and Asia,' where consumers refuse to buy products containing GMOs.[27]

So, those who argue that farmers should have the choice of GM crops are essentially arguing for GM to be the only choice, since in the short to medium term all non-GM and organic crops will be contaminated by transgenes, thereby making them GM too. These GM crops really are their own best marketing technique.

And yet, global organic sales are more than seven times those of GM crops. By 2005, global GM crop sales had stalled at $5 billion a year, whereas those for organic foods and fibre were about $37 billion and climbing, as demand rockets.[28] And that's without mentioning the enormous environmental and health benefits from organic farming, which are valued at £100 per hectare by the UK government itself.[29] Then there's jobs – in Germany alone, organic agriculture has already created 150,000 jobs, compared to only 2,000 jobs in the agbiotech sector.[30]

*Threatening crop biodiversity*

As we have seen, GM crops have already widely contaminated established seed stocks. Also under threat from potential degradation and loss are the indigenous landraces of crops – varieties of seeds bred by farmers over many generations. One example is the important Mexican landraces of maize. In 2003, transgenes were found in 24 per cent of total samples of native corn from nine states, with contamination running from 1.5 per cent to 33.3 per cent.[31] According to Greenpeace, this is in part explained[32] by the 5 million tonnes of US maize imported into the country each year,[33] one-third of which are contaminated with GM varieties from Monsanto.[34]

And for wild populations of agricultural crops – which are valuable gene reservoirs for new useful genetic traits – new mathematical models have shown that they could possibly be obliterated by their GM descendants in less than a decade, which is just an evolutionary blink of an eye.[35]

All this exacerbates an existing problem, which is that agricultural biodiversity is already under immense pressure as multinationals have displaced traditional varieties by buying up all the local seed companies (according to the Soil Association: 'In 1900 there were around 2000 seed companies in North America, now there are less than 200.'[36]) and narrowing the range of crops and varieties on sale around the world. We have already lost 75 per cent of all agricultural genetic diversity that was present in 1900.[37] This means that we have lost a huge amount of genetic material with valuable variations for local conditions.

GM agriculture will only accelerate this loss.

## Other problems with growing GM crops

*Bt crops*

Crops genetically engineered to produce their own pesticides (Bt crops) will lead to the more rapid appearance of pesticide-resistant insects, owing to the continued presence of the pesticide within the plant, which forces a faster rate of adaptation to, and therefore resistance against, these pesticides. This in turn undermines relatively inexpensive pest management systems, which rely on the application of non-GM Bt pesticide straight on to crops.[38]

In fact, a study by entomologists at the Indian Agricultural Research Institute in New Delhi has borne this out, finding that the protection afforded by the GM Bt gene lasts at best for six years. The bollworm developed a '31-fold resistance to the [Bt gene-produced] toxin "Cry1ac" within six generations'.[39]

Even Monsanto – rarely able to resist a sales pitch – acknowledge that insect control through transgenic plants 'may not be desirable in the long term' because it produces resistant strains and 'numerous problems remain ... under actual field conditions'.

Furthermore, the measures put in place to address the problem are being ignored. Farmers are advised to plant no more than 50–80 per cent of their total maize area to Bt crops, leaving the rest as a Bt-free 'refuge'. Yet, nearly 30 per cent of farmers were not following these rather inconvenient guidelines in the year 2000, because, as the Soil Association observes: 'this practical restriction undermines the supposed convenience of Bt crops'.[40]

Recent research has shown that pests also quickly mutate to find Bt a useful food – another fact conveniently edited from the biotech advertising hype – thus potentially increasing the fitness of resistant populations.[41]

A further problem that arises from Bt crops is that they are not sufficiently selective, which means they damage beneficial pest-control and pollinating insects, thus creating a niche for insects that previously were not behaving as pests. In China, the Nanjing Institute of Environmental Sciences reported that GM Bt cotton damages the bollworms' natural parasitic enemies, as well as encouraging other pests.[42] Furthermore, University of California mathematical simulations show that this suppression of natural pest predators often neutralises the Bt gene's positive effect. After a few years of Bt crops, the likely scenario is bigger, more unmanageable pest problems, as beneficial insect biodiversity disappears.[43]

### Creating resistant viruses

As with the creation of resistant insects, a plant genetically engineered to be resistant to a virus will result in the more rapid evolution of new viruses capable of infecting a wider range of plants. In the not too distant future, since viruses can evolve extremely quickly, this, too, will cause even greater problems for farmers.

### Encouraging Fusarium

This is a good one. The weeds may hate Monsanto's Roundup weed killer, used on all GM Roundup Ready crops, but a potentially toxic and sometimes deadly fungus, Fusarium, loves it. Glyphosate – the active ingredient in Roundup – has been shown to increase Fusarium and other microbes, by 'just under 50 scientific papers', says Dr Robert Kremer, a soil scientist at the University of Missouri.

As a sign of how serious this problem is, in Michigan, during 2002, it was estimated that 30–40 per cent of (non-GM) cereal crops (maize is a cereal) were destroyed by such an infestation. This not only calls into question the world's number one weed killer, but also jeopardises Monsanto's flagship line of GM Roundup Ready crops.[44]

### Unpredicted effects

'Genetically engineered crops ... represent a massive uncontrolled experiment whose outcome is inherently unpredictable,' according to Dr Barry Commoner, a senior biologist at the City University of New York.[45] Here are a few more examples to swell the already impressive list of GM misdemeanours:

- In India, the pink bollworm, against which GM Bt cotton is engineered to be resistant, is emerging as a major pest for both GM Bt and non-GM cotton, especially in Andhra Pradesh and Gujarat. The toxin in Bt cotton kills only the green bollworm.[46]
- In the US, tens of thousands of acres of GM Bt cotton malfunctioned in the first year it was planted. In Missouri, some plants shed their bolls (in which the cotton is contained), while others were killed by the very herbicide they were engineered to withstand. In Texas, nearly half the crop did not provide the expected insecticide levels, and 'numerous farmers had problems with germination, uneven growth, lower yield, and other problems'.[47]
- Farmers in the southern US states experienced soy plants splitting in hot weather. This followed higher lignin production by RR soy plants, compared to their non-GM counterparts, which makes for more brittle stems. This results in splitting

and stunting at much greater levels than non-GM soya, and can lead to a 40 per cent yield reduction.[48]

- Herbicide Tolerant GM crops are sometimes more susceptible to disease and may be less vigorous than normal varieties. This is believed to be related to the additional genetic material suppressing the plants' stress responses.[49]

- Potatoes engineered to deter one pest have been found to attract another, demonstrating the virtual impossibility of taking account of all the ecological consequences of making small changes in the biochemistry of an organism.[50]

### Mystery sickness in the Philippines

In March 2004, about a hundred inhabitants from a village in the southern Philippines experienced a strange illness when nearby Bt maize began flowering. Their symptoms included coughing, vomiting, dizziness, headaches, and stomach aches. Dr Terje Traavik, a Norwegian government scientist, suggested a connection between the GM crop and the illness. One of the villagers said: 'There was this really pungent smell that got into our throats. It was like we were breathing in pesticides.' Livestock suffered as well, with five horses dying. One family moved three miles away to stay with relatives and the symptoms disappeared. But the person who rented their home experienced the onset of the same symptoms after their arrival. The mystery illness they all suffered from is not the only such incident involving Bt crops.

In fact, even the natural non-GM form of the Bt toxin has been linked to health problems in those who spray it. But the effects of the GM form of the Bt toxin are likely to be wider ranging, because, unlike the non-GM version of Bt, the active ingredient is switched on all the time. Scientists have already raised questions about its safety.[51]

### Animals and GM crops

It's official, animals do not like GM food – which means they've just overtaken humans in the intelligence league. We're near the bottom, just above pond life.

The Soil Association report lists numerous anecdotes from farmers about how animals avoid GM feed if they can. For example:

- 'If a field contained GM and non-GM maize, cattle would always eat the non-GM first.' – Gale Lush, Nebraska

- 'A neighbour had been growing Pioneer Bt maize. When the cattle were turned out onto the stalks they just wouldn't eat them.' – Gary Smith, Terry, Montana
- 'I saw an advert from a farmer who was looking for non-Bt corn, as he was getting lower milk yields from the cattle that were eating Bt corn.' – Tom Wiley, North Dakota
- 'A captive elk escaped and took up residence in our crops of organic maize and soya. It had total access to the neighbouring fields of GM crops, but never went into them.' – Susan and Mark Fitzgerald, western Minnesota
- 'A student placed two bales of maize in a rodent infested barn. One was Roundup Ready and the other was conventional. Apparently the rodents would not touch the Roundup Ready crop.' – Roger Lansink, Iowa

Furthermore, it is believed that the lack of segregation of GM and non-GM crops has probably 'masked other problems'.[52]

In Iowa, 17 pig breeders reported a sharp decline in pig farrowing (conception) rates, after feeding their pigs Bt maize. Jerry Rosman experienced an 80 per cent fall and four of his neighbours had the same problem. Outwardly, the pigs showed all the signs of being pregnant, but after two to four months the signs disappeared. Laboratory tests eventually revealed that the Bt maize from all five farms had high levels of fusarium mould, suggesting the presence of mycotoxins that can cause pseudo-pregnancies. One of the breeders stopped feeding Bt maize, and farrowing rates returned to normal.[53]

In Germany, 12 cows fed GM maize on a farm in Hesse died in mysterious circumstances. The Robert-Koch Institute in Berlin, which authorised the GM maize, has refused to conduct a full investigation into the deaths. Greenpeace is calling for such an investigation and an immediate ban on Syngenta's GM Bt176 maize.[54] This was the very maize that was being grown on 20,000 ha in Spain in 2003 and that the EU has since asked Spain to withdraw.[55]

And if all that is not enough to convince, read Dr Eva Novotny's excellent report for Scientists for Global Responsibility. The report contains an analysis of Aventis (now Bayer's) feeding trial of Chardon LL GM maize, which the British government has – in its infinite wisdom – approved for commercial growing:

Re-analysis of experiments on chickens and on rats fed Chardon LL GM maize suggest that, contrary to the official conclusions, at least some individual animals do not gain weight as rapidly as they should when given a diet including

GM feed. Furthermore, there appear to be irregularities in the feeding habits of at least some animals given GM feed. In the experiment on chickens, mortality was twice as high among those fed the GM maize as among those fed non-GM maize.

It was a result that Aventis/Bayer claimed was normal and not significant.[56] At what point, I wonder, would increasing human mortality register as significant on the corporate Richter scale?

## GMOs IN FOOD

### Food contamination

*The StarLink fiasco*

Even in the short lifespan of GM foods, and its very limited geographical scope, there have been enough contamination incidents to set the alarm bells ringing.

The worst to date was a nightmare for the biotech PR machine and cost Aventis an immense $1 billion to clear up.[57] In October 2000, in the US, GM StarLink corn, approved only as animal feed, ended up in taco shells and other food products. Farmers and grain elevators hadn't heard about the special handling instructions to keep the corn segregated, which the manufacturer was required to inform farmers about.[58] It led to a massive recall of over 300 food brands.[59] As ABC News observed: 'In Iowa, StarLink corn represented 1% of the total crop, only 1%. It has tainted 50% of the harvest.'[60]

Hundreds of people had allergic reactions to StarLink. According to Marc Rothenburg, Chief Allergist at Cincinnati Children's Hospital, symptoms 'varied from just abdominal pain and diarrhea [and] skin rashes to ... a very small group having very severe life-threatening reactions'.[61] One individual 'experienced immediate respiratory failure after ingesting two taco products'.[62] This resulted in a heart attack and death soon after. At the time, no one realised StarLink was the cause.[63]

Incredibly, the Food and Drug Administration (FDA) had relied entirely on the manufacturer's own safety assessment, which 'concluded that corn ... derived from the new variety [is] not materially different in composition [and] safety ... from corn ... currently on the market [that is, it was Substantially Equivalent]'.[64]

The FDA eventually decided to test StarLink to see if it had caused the allergies. But it trusted the manufacturer to provide for testing the Cry9C protein – the StarLink version of Bt toxin, which is able to resist heat and gastric juices, thus allowing the body more time to overreact and give an allergic reaction. Despite the fact that these were the people who had the most to lose if StarLink was found to be the problem, the FDA did not require verification as to its purity. The Cry9C protein the manufacturer (Aventis) provided was not from StarLink at all, but a synthesised protein substitute derived from E. coli bacteria, which is not the same thing.[65]

Aventis also presented an entirely inadequate document to the Environmental Protection Agency's (EPA) Scientific Advisory Panel,[66] who, astonishingly, maintained their initial judgement that StarLink was a medium likelihood allergen,[67] which is quite shocking considering the life-threatening reactions some people had to StarLink.

A Friends of the Earth report concluded:

The StarLink debacle is a case study in the near total dependence of our regulatory agencies on the 'regulated' biotech and food industries. If industry chooses to submit faulty, unpublishable studies, it does so without consequence. … perhaps FDA chose the easy course of reliance on Aventis to avoid making trouble for the industry it openly promotes.[68]

The StarLink gene is obviously a committed troublemaker, because it has continued to pop up ever since – in food aid sent to Bolivia in 2000,[69] in a 2002 shipment to Japan,[70] in food aid sent to Central America in 2002 and 2004,[71] and in other US corn exports, to the extent that it has even contaminated native Mexican maize varieties.[72] Even in 2003, three years after the event, it was still showing up in more than 1 per cent of samples submitted in the US.[73]

One US corn farmer and GM seed salesman put it this way: 'you guys created this monster; you clean it up. I have learned my lesson. No more GMO crops on this farm – ever.'[74]

### The scandal of Syngenta's Bt10

In March 2005, Syngenta admitted that it had accidentally produced and disseminated – between 2001 and 2004 – 'several hundred tonnes' of a corn called Bt10, which was unapproved in the EU[75] and sold as an approved corn, Bt11. About 15,000 ha was planted in the US,[76] which means that about 150,000 tonnes went into the food chain.[77] Syngenta representatives wouldn't say which countries had inadvertently received the wrong seed.[78]

Adrian Bebb, from Friends of the Earth, said:

This is an industry out of control. For four years Syngenta failed to notice that they were selling farmers an unapproved genetically modified seed. How are consumers and farmers supposed to trust them ... in the future? This case makes a complete mockery of the US regulatory system for GM crops. To make matters worse the US Government has known about this accident for [4] months and together with Syngenta decided to keep it a secret until now.[79]

It was *Nature* that first reported the story on 22 March 2005. There then followed, according to GM-free Cymru: '... a carefully coordinated "damage limitation campaign" involving statements to the press which were extremely economical with the truth'. DEFRA's statement, rushed out a day later and worded to protect the biotech industry, was, said GM-free Cymru, 'evasive and dishonest' and 'contained lies which have still not been officially admitted or corrected'. The NGO also said that the official UK and European response was 'incompetent and complacent' – with the European Commission (EC) taking 10 days to act.[80] Furthermore, the EC and the UK's FSA gave assurances that all the Bt10 became animal feed, thus making it fairly harmless, and yet Syngenta admitted in an email to DEFRA (obtained by GM-free Cymru, under the Freedom of Information Act) that the five Bt10 lines were a type of corn used in many processed foods.[81]

The official line was that Bt10 was 'safe', as it was considered almost identical to Bt11. However, according to Dr Jack Heinemann, Institute of Gene Ecology, University of Canterbury:

The Syngenta documents ... indicate that there are additional and possibly substantial differences between Bt10 and Bt11 ... there were differences in the profiles of PAT and Cry1Ab proteins and thus there may be other undetected differences ... The PAT levels in Bt10 appear to be much higher than in Bt11. ... [which] could indicate a higher overall stability of the PAT protein, and possibly Cry1AB protein ... produced by Bt10. [Thus] their potential to be allergens may be underestimated by studies using Bt11 as the source.[82]

Syngenta itself delayed releasing the complete information on Bt10. This was particularly evidenced by the fact that neither Syngenta nor the EPA mentioned the existence of a GM antibiotic resistant marker (of a type the EU has been advised to phase out[83]) when they first admitted the scandal. Syngenta was similarly reluctant to cooperate with another GM contamination incident in 2000, when it supplied 5.6 tonnes of GM-contaminated maize seed, which was illegally transported to New Zealand and then sown as non-GM seed.[84]

Once again, the US is shooting itself in the foot, not only with its GM policy, but also in the way it handles such situations. This is clear from the fact that Japanese importers have almost halted new purchases of US corn for fear of Bt10 contamination, with some Japanese purchasers having apparently moved to non-US corn.[85] Furthermore, in mid April 2005, Europe finally banned US maize imports until they were proven Bt10-free.[86]

### Illegal GM rice enters food chain in China

In April 2005, Greenpeace discovered unauthorised GM Bt rice that had been sold and grown unlawfully for the previous two years in the Chinese province of Hubei. An estimated 950 to 1,200 tons of the rice entered the food chain after the 2004 harvest, with the risk of up to 13,500 tons entering the food chain in 2005 unless remedial action was taken quickly. The rice may also have contaminated China's rice exports.[87] Chinese merchants said the supplies were bought from the Huazhong Agriculture University in Wuhan, which specialices in GM rice research.[88] Furthermore, the *Guardian* reported that the seeds were sold unlawfully on the internet to Chinese growers.[89] 'The GE industry is out of control,' said Greenpeace's Sze Pang Cheung:

A small group of rogue scientists have taken the world's most important staple food crop into their own hands and are subjecting the Chinese public to a totally unacceptable experiment. We're calling on the Chinese government ... to recall the unapproved GE rice from the fields and from the food chain...[90]

The Chinese government responded by prohibiting all media coverage of the debacle.[91]

### Other contamination

There have also been numerous other smaller incidents, as well as ongoing contamination. One of the earliest was in 1997, when Greenpeace revealed that conventional soybeans imported into Britain contained GM soybeans.[92] Other examples of contamination or potential contamination include: the unauthorised sale of GM research piglets for food;[93] the sale of toxic corn by the USDA, which will have found its way into 'food and feed channels';[94] and the appearance of GM pork, from stolen pigs, in a sausage served at a funeral dinner – the pigs had been genetically engineered to develop a disorder similar to diabetic blindness in humans.[95]

Many believe agricultural biotechnology is simply not controllable.

## Biopharmaceutical crops

And if that all sounds like a nightmare... you ain't seen nothing yet.

Biopharming is the genetic engineering of drugs and chemicals into crops – the majority of biopharmaceuticals being incorporated into maize, along with such other crops as rice, soy, and tobacco. In theory, GM plants will produce chemicals cheaper than a factory, although even that is now in doubt.[96] So, soon near you, an exciting new range of contaminants in your food, including: vaccines and contraceptives; plus drugs and chemicals to induce abortions, create blood clots, produce industrial enzymes, and propagate allergenic enzymes. These are just some of the chemicals that could be engineered into crops. No health risk left unturned.

No pharmed human drug has yet come on to the market, and only a few companies have got past test plantings, but it is projected that 10 per cent of the US maize crop will be devoted to biopharm products by 2010.[97] In 2002, there were only about 300 acres planted with pharm-crops in the US, yet despite such a minuscule area, a study by the Union for Concerned Scientists warns that pharm crops could already be poisoning ostensibly GM-free food crops by cross pollination and thus contaminating the food supply.[98] We have no way of knowing owing to the extreme secrecy of the 300 plus field trials across the US since 1991.[99] However, there have been a couple of well-documented and nerve-racking near misses.

### 'Pharmageddon' and other stories

In November 2002, the US government ordered biotech company ProdiGene to destroy 500,000 bushels of soybeans in Nebraska. Bound for human consumption, they were contaminated with GM corn, containing a pig vaccine neither approved for human consumption nor tested for human health effects.[100] Professor Norman Ellstrand, a world-renowned corn geneticist at the University of California, said the US government had been lucky. 'What if the GM corn had come up inside a corn field, instead of a soybean field? It could have cross pollinated and you'd have no idea where it was.'[101]

The day after, the US government disclosed that ProdiGene had, in fact, done the same thing in Iowa two months earlier. ProdiGene failed to follow government regulations for the containment of the crop. Fearing that pollen from corn, again not approved for human consumption, may have spread, the USDA ordered 155 acres of nearby corn to be incinerated.

But why the delay in telling the public – is the USDA hiding similar problems? And if the Nebraska contamination had not been discovered by the media, would the USDA have disclosed the Iowa incident?[102]

### The aftermath

In a none too subtle sign of where his loyalties lie, President George Bush honoured ProdiGene's president – in the same month that the Iowa contamination occurred – by personally appointing him to America's Board on International Food and Agriculture Development, the sole adviser to the US Agency for International Development (USAID).[103] Furthermore, rather than a tough crackdown after the Nebraska incident, the US government kindly gave ProdiGene as much as $500,000 (of taxpayers' money) in financial assistance to help pay for the clear-up, which cost the company $3.5 million in costs and $250,000 in fines. The USDA never disclosed this – it was only uncovered later by the Center for Science in the Public Interest.[104]

Jean Halloran of Consumers Union had this to say on pharmaceuticals:

The practical aspects of trying to keep these pharmaceutical plants separate from the regular food plants is an insurmountable problem. It just can't be done. It can't be done because of the fallibility of human beings. It can't be done because you can't control pollen flow. It can't be done because you can't control mother nature that way. And if you can't control mother nature and fallible human beings, we've come to the conclusion that you shouldn't try.[105]

Jane Rissler, senior scientist with the Union of Concerned Scientists, observed:

If a company cannot be relied upon to perform such a simple task as to keep pharm corn out of soybeans, how can it be trusted in the far more complicated process of keeping drugs out of corn flakes?[106]

Opposition to biopharming is increasing all the time. The Genetically Engineered Food Alert coalition is calling on the USDA to prohibit open-air cultivation of all biopharm crops.[107] The food industry is nervous too; it is pressing the biotech industry to make pharmaceuticals only from non-food crops such as tobacco. Even the pro-GM *Nature Biotechnology* journal published a damning editorial in February 2004 on biopharmaceuticals in food or animal feed crops, fearing that the practice will muddy the pitch for the whole GM industry.[108]

### The dangers of using viral genes and promoters

Probably the greatest threat from genetically altered crops is the insertion of modified virus and insect virus genes into crops. It has been shown in the laboratory that genetic recombination will create highly virulent new viruses from such constructions.

These are the sobering words of Dr Joseph Cummins, Professor Emeritus in Genetics from the University of West-Ontario.[109] These potentially virulent new viruses may in turn cause famine through new crop diseases. Or they might cause powerful new animal and human diseases. The question is – do we want to risk unleashing viral epidemics just so that Monsanto et al. can make even more profits?

Dr Cummins also observes that a promoter (which allows a GM gene to be 'switched on' in the host genome) can have 'the same impact as a heavy dose of gamma radiation', by creating a 'hotspot' in the DNA, which is a place where a whole section of DNA can become unstable.[110] 'This may cause breaks in the strand or exchanges of genes with other chromosomes,' observes Dr. Terje Traavik.[111]

Furthermore, in unpublished research from 2004, Dr Traavik, a geneticist and adviser to the Norwegian government, and Director of the Norwegian Institute for Gene Ecology, found the Cauliflower mosaic virus (CaMV) promoter (used in the vast majority of GM crops to date) intact in rat tissues after just a single meal. In a separate but related study, it was also confirmed to be active in human cells, including in cells in the stomach, intestinal lymph, kidneys, liver, and spleen. These experiments are disturbing because they contradict one of the biotech industry's arguments that claim to prove that CaMV is safe, which is that the CaMV won't transmit from food to gut bacteria or internal organs. Another issue that is particularly worrying is the possibility that, as Dr Traavik observes, the CaMV promoter can turn on ancient but dormant viruses embedded into human (or any other organisms') DNA.[112]

### Antibiotic resistance

An Antibiotic Resistant Marker (ARM) gene is generally used to indicate which cells have obtained the new genetic material. The concern is that, when GM food is eaten, the ARM genes will pass into the bacteria of the host's digestive system. Because the ARM gene is resistant to antibiotics, if it moves from species to species, new antibiotic resistant diseases could result.[113]

As always, the biotech companies have reassured us this won't happen. But, according to Dr Michael Antoniou:

There is a whole slew of different antibiotic resistant genes that are being used in GM crops in their production in the laboratory. They stay in the final crop. ... Bacteria in the gut are going to take up genes that will make them resistant to potentially therapeutic antibiotics. The possibility is that [for] someone who picked up the antibiotic resistance through food and then fell ill ... a medical antibiotic might not be effective.[114]

Bodies calling for ARM genes to be phased out include the World Health Organization, the House of Lords, the American Medical Association, and even the Royal Society.[115] According to the FDA website, antibiotic resistant infections 'increase risk of death, and are often associated with prolonged hospital stays, and sometimes complications. These might necessitate removing part of a ravaged lung, or replacing a damaged heart valve.'[116] In the UK, antibiotic resistant bacteria kill more than 3,500 people a year.[117]

### New food toxins and allergens

Dr Andrew Chesson, Vice Chairman of European Commission Scientific Committee on Animal Nutrition, and formerly an ardent advocate of food biotechnology, predicts that 'Potentially disastrous effects may come from undetected harmful substances in genetically modified foods.'[118]

And Samuel Epstein, MD, Professor of Environmental Medicine at the University of Illinois School of Public Health and Chairman of the Cancer Prevention Coalition, observes that 'Manipulating DNA creates a new substance and it may not behave in the same ways as the original version. And existing tests, which only detect already-known toxins, may not reveal man-made ones. *We simply do not know what we are doing*' (my emphasis).[119]

### *L-tryptophan*

In 1989, a bacteria genetically modified to produce large amounts of the food supplement, L-tryptophan, yielded impressively toxic contaminants that killed 37 people, partially paralysed 1,500, and temporarily disabled 5,000 in the US.[120] The epidemic it spawned – Eosinophilia-Myalgia Syndrome (EMS) – had some horrendous symptoms. One man described his legs becoming 'as big as a telephone pole. They split and water oozed from them. No amount of medicine … calmed the pain.'[121] Those who suffered most experienced an 'ascending paralysis, in which a person loses nerve control of the feet followed by the legs, then bowels and lungs, finally requiring a respirator in order to breathe'.[122]

The biotech industry's response was that this resulted merely from quality control failure.[123] The company itself has rather conveniently destroyed the bacterium, making further research impossible.[124] But evidence shows that genetic engineering itself caused the bacteria to generate unexpected toxic derivatives.[125] Only the GM version of L-tryptophan caused these terrible problems.[126]

The FDA – intent on protecting the infant biotech industry – withheld[127] and covered up vital information,[128] propounded discredited theories,[129] and generally acted in an improper manner. It never admitted that the epidemic was caused by genetic engineering and spun the episode to suit its own agenda to justify that food supplements were dangerous and therefore needed regulation.[130] In a rather telling gesture, Showa Denko, the company involved, extended out-of-court settlements to some of the victims – a massive $2 billion to more than 2,000 people.[131]

According to Dr Michael Antoniou:

If this GM product was produced today, it would have been passed as substantially equivalent to the non-GM version, because it was greater than 99% pure, thus avoiding the various trials that are required to detect novel toxins of the type that was produced. Furthermore, the suspected novel toxin, which caused all the problems, was present at less than 0.1% in the final product, a level at which the contaminants would not have to be identified in the UK. Therefore, by these criteria, the toxin present in the tryptophan would not have attracted any attention or concern, and so the tragedy would still have occurred to this day in the UK.[132]

*Other causes*

On a less dramatic scale, but no less worrying:

The unexpected production of toxic substances has now been observed in genetically engineered bacteria, yeast, plants, and animals, with the problems remaining undetected until a major health hazard has arisen. Moreover, [GMOs] may produce an immediate effect or it could take years for full toxicity to come to light. – Dr Michael Antoniou[133]

The problem is that few countries keep food allergy records. Luckily, however, the UK's York Nutritional Laboratory does. In March 1999, its scientists observed that during the previous year soy allergies had rocketed by 50 per cent. The range of chronic illnesses included irritable bowel syndrome, digestive problems, and skin complaints such as acne and eczema, together with 'neurological problems, with chronic fatigue syndrome, headaches, and lethargy'. Most of the soy in the UK was from the US and so contained GM. Dr Antoniou observed that 'At the moment no allergy tests are carried out before GM foods are marketed.'[134] And, as a *Washington Post* article from 1999 reports, there is 'no widely accepted way to predict a new food's potential to cause an allergy. ... With no formal [FDA] guidelines in place, it's largely up to the industry to decide whether and how to test for the allergy potential of new food not already on the FDA's "must test" list.'[135] The FDA's microbiologist Dr Louis Pribyl is even less reassuring: 'the only definitive test for allergies is human consumption by affected peoples'.[136]

### Bt toxins

GM Bt crops are engineered to contain Bt-toxins from soil bacteria, which in theory make the plants pest resistant. In other words, these plants have built-in pesticides, so, by definition, we are eating pesticides when we eat GM Bt crops. According to Dr Michael Hansen of the Consumers Union: 'There is increasing evidence ... that the various Bt endotoxins – including those from [GM corn], cotton, and potatoes – may have adverse effects on the immune system and/or may be human allergens.'[137] Interestingly, Bt is very similar to Ba – *Bacillus anthracis*, or plain old anthrax.[138]

The EPA asserts that the Bt toxin is supposed to be destroyed in the stomach, but that claim is based on the biotech companies' own research. In Monsanto's experiments to show how quickly its Cry1Ab (Bt corn protein) broke down, the company used a much stronger acid and vast amounts more enzyme than would occur in the stomach to digest tiny amounts of Cry1Ab. Critics assert that this process was unrealistic and fashioned to eradicate the protein as swiftly as possible.[139] According to author Jeffrey M. Smith: 'Neither the EPA nor FDA has established standards for these tests – they have thus far accepted whatever procedures and conclusions the biotech companies have given them.'[140]

## Possible carcinogenic effects

Given the huge complexity of genetic coding, even in very simple organisms such as bacteria, no one can possibly predict the overall, long-term effect of GM foods on the health of those who eat them. Many scientists worry about the carcinogenic effects of GM foods.

Professor Samuel Epstein observes that 'Genetic modification of food is a dangerous game ... I'm sure there will be a significant increase in deaths from certain types of cancer. If that is the only adverse effect we will have been lucky.'[141]

Dr Ewen and Dr Pusztai's 1999 ten-day study on male rats fed GM potatoes is one of the few studies on the effect of GM on the gut. It observed 'a powerful effect on the lining of the gut (stomach, small bowel, colon and rectum) leading to proliferation of cells', according to Dr Ewen, a consultant histopathologist, and a leading expert in tissue diseases. His concern is that 'the proliferation factor could then act on any polyp present in the colon, which is regularly accessed by GM material, and drastically accelerate the development of cancer in susceptible persons'. Dr Ewen observes that most people in countries with a genetic disposition towards colon cancer (for example, NE Scotland, North America, and New Zealand) will have these polyps by the time they reach 70. Interestingly, these effects were not noticed in either of the two control groups of rats. Dr Ewen considers that 'the CaMV was the realistic causative agent', although other scientists disagree with this conclusion. What is certain, though, is that these findings are worrying and further research is needed to identify the causative effect. As Dr Ewen comments: 'This question remains – why should I have to eat, for the rest of my life, a highly active virus fragment that has not been adequately tested in animals?'[142]

With no labelling or monitoring of GM foods, in places like the US, the extra cancers caused by GMOs will be lost in the already high cancer levels. In other words, no one will be able to say whether or not they were from GMOs, which is why the biotech lobby so vehemently opposes GM labelling.

## Possible effects on babies and children

Other delights the gene crunchers have created for us include the effects on our children. Dr Vyvyan Howard, expert in infant toxicopathology at Liverpool University Hospital, observes that 'Swapping genes between organisms can produce unknown toxic effects and allergies that are most likely to affect children.'[143] Michael

Meacher has cited a Royal Society report, which said: '... that any baby food containing GM products could lead to a dramatic rise in allergies, and unexpected shifts in oestrogen levels in GM soya-based infant feed might affect sexual development in children'.[144]

### More risky than traditional foods

It is interesting to hear what the medical establishment has to say about this. The British Medical Association (BMA), which represents over 80 per cent of practising doctors in the UK, is hardly a bunch of dreadlocked radicals. Yet, in 2002, it said: 'There has not yet been a robust and thorough search into the potentially harmful effects of GM foodstuffs on human health.' 'Insufficient care' had been taken over public health.[145]

Dr E.J. Matthews of the FDA's Toxicology Group warned that 'genetically modified plants could ... contain unexpected high concentrations of plant toxicants'. He cautioned that some of these toxicants could 'be uniquely different chemicals that are usually expressed in unrelated plants'.[146]

A statement by 21 scientists, including Professors Richard Lacey (microbiologist and Professor of Food Safety at Leeds University) and Brian Goodwin (Professor of Biology at the Schumacher College), said: 'With genetic engineering, familiar foods could become metabolically dangerous or even toxic.'[147]

GM crops are not just uniquely risky in their own right, though. They are also likely to contain higher pesticide residues. The Conservative government in the UK made sure of that. In 1997, in a rather touching sign of its loyalty to Big Business right up to the bitter end, it gave Monsanto a parting gift by quietly raising the permitted levels of glyphosate (active ingredient of Roundup) in soybeans destined for food by 200 times. This was to prevent Monsanto's Roundup Ready GM crops from containing illegally high levels of the chemical, thus clearing away legislation that hindered the biotech industry.[148]

### MORE PROBLEMS FOR FARMERS

#### The national farm economy

The Soil Association report found that GM crops were an economic disaster in the US. Rather than improving competitiveness, they are a major burden on the agricultural economy. They have cost the US an estimated $12 billion since 1999[149] (that's enough to restore the sight to 480 million adults suffering from cataracts in the developing countries[150]) in lower crop prices, lost sales, in farm subsidies, and product recalls owing to transgenic contamination.[151] And in Canada, an August 2003 government paper, marked 'secret', warns that GM crop commercialisation has put the country's multibillion-dollar agrifood industry at risk.[152]

*Loss of international trade*

According to the Soil Association report: 'The most dramatic result of GM crops [in the US and Canada] has been the complete collapse of major export markets' for the crops where GM varieties are grown. This affects both the GM and non-GM trade in those crops.[153] George Bush has responded by trying to force GM crops on other countries. The American Corn Growers Association (ACGA), and other organisations, observe that this is 'Hardly a consumer-oriented approach' and that surely it would be more sensible to supply what the market wants.[154] This seems to be backed up by news that, at the end of 2004, for the first time in nearly 50 years, the US had an agricultural trade deficit. As recently as 2001, there had been a $13.6 billion trade surplus. It is no coincidence that this follows nine years of the US as the world's largest GM crop grower.[155]

*Increased US farm subsidies*

The Soil Association report observed that

US farm subsidies were meant to have fallen over the last few years. Instead, they rose dramatically, paralleling the growth in GM crops. The lost export trade as a result of GM crops is thought to have caused a fall in farm prices and hence a need for increased government subsidies, estimated at an extra $3–5 billion annually.[156]

Generous to a fault, George Bush has supported the main GM crops with the new subsidies, with soy and maize receiving 50 per cent of all subsidies. In fact, soy subsidies only started in 1998, shortly after the commercialisation of GM crops.[157] Clearly, this is unfair support for GM crops, which will help mask their economic failings. Once again, the books are being cooked.

### Reduction of good non-GM varieties

The Soil Association report found that 'Over the last few years, as GM varieties were introduced, the availability of good non-GM varieties

... has been significantly reduced.' This too will skew the relative performances of GM and non-GM crops firmly in the direction of the latter. This is hardly likely to be a coincidence, given that the GM corporations are also the main seed sellers.[158] 'There is no choice. It is so difficult to find guaranteed non-GM soya, and you end up with poorer varieties.' – Gale Lush, Chairman of American Corn Growers Foundation (ACGF) and GM soy farmer, Nebraska.[159]

## A social and environmental nightmare in Argentina

As Daniela Montalto, an Argentinian GM campaigner, put it, growing GM soy has been an 'environmental and social nightmare in Argentina'.[160]

In the last decade, the amount of soy grown in Argentina has nearly tripled,[161] covering 43 per cent of farmland in 2002; 90 per cent of it is GM. It has replaced a once-mixed farming system,[162] so that production of many staples, including milk, rice, maize, potatoes, and lentils, has fallen sharply. In a country used to producing ten times more food than the population needed,[163] once-plentiful exports of beef and wheat have dried up.[164] According to a recent report, more than 250,000 Argentinian children are suffering from malnutrition because the cheap, local crops they once ate are no longer available.[165] 'As Argentina's GM soy exports increase, so do hunger, marginalisation and destitution in this once plentiful land,' report Ann Scholl and Facundo Arrizabalaga in an article entitled 'Argentina: The catastrophe of GM soya'.

They continue:

The countryside is being left empty as the farm workers' role in nurturing the land and crops is displaced by aeroplanes and agribusiness infrastructure ... Migration to the cities has risen at an alarming rate: 300,000 farmers have deserted the countryside and more than 500 villages have been abandoned, or are on the road to disappearance. ... As a consequence, the *villas miseria* on the outskirts of the cities are mushrooming with the arriving unemployed agricultural workers.[166]

Dr Mae-Wan Ho, Director of the Institute of Science in Society (ISIS), reports that

Weeds have multiplied, as resistance to glyphosate (the herbicide used with RR soya) soared. The herbicide has had to be applied more frequently and at higher concentrations. Toxic older herbicides, such as 2,4 D and Paraquat, banned in many countries are back in use. ... Aeroplanes are used to spray herbicides on RR soya, subjecting local populations to tremendous health risks.[167]

An article in the *New Scientist* describes how spray drift from a nearby GM soy farm wafted into the village of Colonia Loma Senes.

'The poison got blown onto our plots and into our houses,' recalls local farmer Sandoval Filemon. 'The children's bare legs came out in rashes.' The following morning the village awoke to a scene of desolation. 'Almost all of our crops were badly damaged. I couldn't believe my eyes,' says Sandoval's wife, Eugenia. Over the next few days and weeks chickens and pigs died, and sows and nanny goats gave birth to dead or deformed young. Months later banana trees were deformed and stunted and were still not bearing edible fruit. ... A month later, agronomists from the nearby National University of Formosa visited the scene and confirmed ... that the neighbouring farmers ... had carelessly drenched the land – and nearby Colonia Loma Senes – with a mixture of powerful herbicides.

This is by no means an isolated case.

And it is not the only health problem associated with soy. According to Jorge Rulli, one of Argentina's leading agronomists:

We have been assigned a role in the world as a producer of soya and in many ways we are now a laboratory. We are seeing all kinds of things due to toxicity: precocious sexual development, early pregnancies, and at the same time, stunted growth.[168]

According to an article in *New Scientist*:

Agricultural economics consultant [Dr] Charles Benbrook reported that Roundup Ready soya growers in Argentina were using more than twice as much herbicide as conventional soya farmers, largely because of unexpected problems with tolerant weeds.

University of Buenos Aires agro-ecologist Walter Pengue estimates glyphosate consumption reached 150 million litres in 2003, up from just 13.9 million litres in 1997.[169]

According to another article:

Floods without precedence are taking place as forests are cut down to make way for [GM] soya crops. In the high-mountain provinces of Salta and Juyuy, on the border of Bolivia, the subtropical Yungas region is being deforested to make space for [GM] soya plantations. Greenpeace has warned that in five years, the ancient cloud forest will be extinct.[170]

This forest contains a large selection of animals, plants, and birds, including monkeys, pumas, jaguars, and over half of all Argentinian bird species. Furthermore, hundreds of thousands of people live in this forest – one of the three largest in South America – and this extensive clearance has sparked violence and protest by families dependent on its diversity for their livelihood.[171]

And on top of this, 'Farmers' homes have been bulldozed, paramilitaries have tortured MOCASE [a farmer's movement] members, who also suffer political persecution.'[172]

### GM crops are uninsurable

The biotech companies want it so much their own way that they will accept no liability when their technology goes out of control, as it so frequently does. In Britain, the insurance companies, always a good judge of financial risk, realise what an infinite liability GM crops are – to the extent that the top five companies declared in 2003 that they will not insure them, relegating them instead to the box marked 'Thalidomide, Asbestos, and Acts of Terrorism'.[173] So where does this leave farmers? Up a certain creek without a paddle. In other words, the UK law provides neither protection for those growing GM crops, nor for those who don't but are affected by GM contamination,[174] which means that when it all starts to unravel, the financial buck will stop at the farmer.

In the US, the insurance industry similarly says that GM may be uninsurable, because the FDA has no regulations on GMOs. 'When it comes to a drug or medical device, what underwriters look to as most important is FDA oversight,' said Thomas Greany, Senior Vice President at Marsh, a risk-management firm. This gives the industry a framework to deal with adverse outcomes. Robert Hartwig, the chief economist for the Insurance Information Institute, an industry trade association, observes that 'Genetically modified foods are among the riskiest of all possible insurance exposures that we have today.'[175] The industry is also concerned that juries may not be predictable, and they are uncertain about which GM risks are insurable.[176] US insurers are worried about expensive class-action suits against biotech companies. Currently, the only coverage biotech companies can get is limited and expensive, with big and growing gaps between the possible payouts and insurance coverage.[177]

David Moeller, attorney at the Farmers' Legal Action Group (FLAG), St Paul, Minnesota, says:

Farmers assessing the costs and the benefits of growing GMO crops should base their decisions not only on production costs and expected yields, but also on the legal liability they may incur by planting, growing, and marketing GMO crops.[178]

## Biopiracy and patents

Biopiracy (the industry prefers the term 'bio-prospecting', in the great tradition of American euphemisms like 'collateral damage') is well illustrated by the Basmati rice patent, which was granted to a Texan company RiceTec in 1997. The company's rice, according to the Five Year Freeze report, had been 'derived from Indian Basmati crossed with other native varieties developed by Asian farmers over many years. RiceTec's patent claims this derivation as an invention. By doing so, RiceTec threatens the export market of the very farmers who have grown and bred the rice varieties ... over many generations.' According to Shivani Chaudhry, of the Research Foundation for Science, Technology, and Ecology, in India: 'The patent is a direct appropriation of traditional knowledge of Indian farmers. It reduces years of informal research, breeding and innovation to a pirated product patent.'[179] As a result, biodiversity in developing countries is pirated for the benefit of the biotech companies rather than for the public good.

Author Luke Anderson observes: 'It is extraordinary that a company can make a single genetic alteration to a plant, and claim private ownership to it as their invention.'[180] And: 'Traditional knowledge systems, and the people who have cultivated biodiversity over thousands of years, count for less in patent law than routine laboratory procedures.'[181]

The poorly legislated developing countries are the ideal place for the seed scouts of the unscrupulous GM corporations to go in search of cheap seeds with potential commercial applications. Of particular interest are the rainforests and the large gene banks in the developing countries. According to Christian Aid, biopiracy is swindling developing countries out of $4.5 billion every year.[182] When biotech companies do pay for genetic material – an event so rare as to be verging on the miraculous – the terms of trade are totally inadequate and payments derisory.[183]

Now, six multinationals control 70 per cent of patents on the five main food crops of rice, wheat, maize, soy, and sorghum.[184]

Syngenta has been particularly lively in this arena. For example, it has acquired exclusive commercial control over Golden rice, tying it up in nearly 70 patents.[185]

In 2002, Syngenta – like an alcoholic unable to resist another drink – was at it again in India, secretly planning to buy the rights of research on 23,000 publicly owned and funded rare varieties of rice from the world's second biggest collection of rice plasm at Indira Gandhi Agricultural University. In the end, after a public furore and accusations of corruption and biopiracy, Syngenta bowed out.[186] The company is, however, still working in 'collaboration' with many other institutions in India.[187]

Dr John M. Conley, Professor of Law at the University of North Carolina, charges that biopiracy and gene patenting blocks progress.

People who look at this from an economic perspective say it seems inefficient and counterproductive to let others monopolize genes, proteins, etc. before the full range of uses you can put them to can be known. It seems like you will have a blocking effect on future progress. Many other people are saying such patents are just wrong and ought to be limited.

Blocking progress is clearly a virtue when it suits the biotech lobby, and yet those who have the temerity to oppose GM are routinely blasted as Luddites.[188]

Professor Wangari Mathai, of the Green Belt Movement Kenya, the first African woman to win the Nobel Peace Prize, puts it more strongly:

History has many records of crimes against humanity, which were also justified by dominant commercial interests and governments of the day... Today, patenting of life forms and the genetic engineering which it stimulates, is being justified on the grounds that it will benefit society, especially the poor, by providing better and more food and medicine. But in fact, by monopolising the 'raw' biological materials, the development of other options is deliberately blocked.[189]

Even the usually pro-GM Professor Borlaug, the father of the Green Revolution, spoke out against patents, saying: 'We battled against patenting. I and the late Glen Anderson (of International Wheat and Maize Research Institute) went on record in India ... and always stood for free exchange of germplasm.' Talking about the US insistence on patents, he said: 'God help us if that were to happen, we would all starve.'[190]

## Genetic Use Restriction Technologies (GURTs)

*Terminator*

The Terminator Technology is a group of genes that confers sterility on any plant it is engineered into. The biotech industry would have us believe that it is desirable to stop spilt GM seed from germinating and spreading. However, its critics believe that it genetically engineers sterility into crop plants so that farmers can't replant seeds, for no other purpose than to protect and enforce corporate patents on GM seeds. Control of seed production will move from the farmer's field to corporate headquarters, and farmers will become wholly dependent on corporations for seeds. As the *New York Times* put it: 'The Terminator will allow companies like Monsanto to privatise one of the last great commons in nature – the genetics of the crop plants that civilisation has developed over the past 10,000 years.'[191]

Terminator is also of zero use to farmers and consumers, and poses extraordinary threats. For example, what if these genes contaminate wild species – how many plants will become extinct? Even the UN Food and Agriculture Organization (FAO) has stated that 'the terminator seeds are generally unethical'.

What is particularly disturbing is that the Terminator Technology was developed with taxpayers' money by the US Department of Agriculture (and a seed company that Monsanto was in the process of buying).[192] And despite being suspended by a worldwide agreement in 1999, because it is so damaging to the world's poor, AstraZeneca and Novartis are still pursuing it, and patents are still being granted.[193] Monsanto says it won't use Terminator, but still holds the patents just in case.[194] And most chilling of all, the USDA stands to make 5 per cent royalties on all net sales and has applied for patents in at least 78 countries. USDA spokesman Willard Phelps explained that the USDA wants this technology to be 'widely licensed and made expeditiously available to many seed companies'. The goal, he said, is 'to increase the value of proprietary seed owned by US seed companies and to open up new markets in Second and Third World countries'.[195] Harry Collins of Delta and Pine Land, eighth biggest seed producer in the world,[196] admitted in January 2000: 'We never really slowed down [on Terminator]. We're on target, moving ahead to commercialise it.'[197] As the Rural Advancement Foundation International (RAFI) point out: 'Unless it is banned by governments, Terminator is going to happen, and probably sooner rather than later.'[198]

The question is – why has Terminator Technology not been outlawed? Why is the public good again being overridden by the wishes of Big Business? Why do our governments seem determined only to listen to the voices of the rich and powerful?

*Traitor*

This is a variation on the theme of Terminator. It is designed so that the spraying of proprietary chemicals is required to turn on certain traits, for example pharmaceutical production, or to turn off negative characteristics like allergen production.[199] Novartis has a particularly cynical version of this, whereby the plant will essentially only survive when sprayed with a Novartis chemical.[200] In other words, the plant is turned into a chemical junky, completely dependent on Novartis to keep it alive, for no other reason than to make even more profits for the company.

The most alarming Traitor plants are those with weakened immune systems, developed to research the effects of pathogens. What sort of disaster would ensue if these traits were to contaminate major food crops? It's a chilling thought, given the biotech industry's record on contamination.[201]

### Farmers sued for patent infringement

In a sneak preview of what a corporate-controlled future will look like, unless we oppose it at every opportunity, we need look no further than Canada. Forget the customer is right. Or even Service-with-a-smile. Welcome to Service-with-a-lawsuit.

Percy Schmeiser, a 73-year-old farmer from western Canada, was a non-GM farmer whose fields became contaminated with unwanted GM pollen. Although he tried to contain the GM contamination, at considerable cost to himself, by buying new seed (he used to save his seed), 20 per cent of his harvest was contaminated. Unhappily for him, the Monsanto man paid him a visit and found the (unwanted) GM plants on his farm. If Monsanto's GM rapeseed is grown without a licence fee of $15 (Canadian) per acre and a correct contract, legal proceedings for infringing its patent follow, even if the GM plants resulted not from planting, but contamination.[202]

Monsanto took Percy to court, where, incredibly, it was ruled that he was liable, regardless of the means of contamination – deliberate or accidental. In other words, under patent law, all contaminated plants – as if by some convenient corporate magic spell – suddenly become the property of Monsanto, and farmers have no rights. The lawsuit cost Percy 25 years of seed research and his life savings ($600,000 Canadian).[203] Incredibly, though, according to the *Ecologist* magazine: 'The corporation later admitted that Schmeiser had not obtained the seeds illegally, but said that wasn't important.'[204]

Percy believes 'Monsanto lost its right to exclusivity [its patent] when it lost control of its invention.'[205] Brian Helweil, agricultural expert with the Worldwatch Institute, reckons: 'It's an absurd situation, akin to someone dumping junk on your land and then accusing you of stealing it.'[206]

Percy has filed a countersuit against the company. He believes Monsanto is being less than honest about the percentage of GM in the samples it took, and he has raised doubts about the samples it used, based on the fact that his were clearly identifiable for being 'full of chaff' (seed husks), whereas those used in court were clean.[207]

Percy has also appealed to the Canadian Supreme Court. And now he travels the world, using his story to explain how farmers' rights all over the world are being undermined by the biotech industry. Monsanto is suing or threatening 1,500 North American farmers.[208] One Mississippi farmer has already been ordered to pay Monsanto $780,000 for 'patent infringement'.[209] Many farmers have settled out of court, which involves signing a gagging clause, whereby Monsanto can publicise the 'villains', but farmers are legally bound to remain silent on their cases.[210]

Monsanto has never lost a case, which is hardly surprising, given its ferocious arsenal to beat farmers into submission – $10 million a year and a staff of 75. Monsanto rather grotesquely boasts a list of victories in its 'Seed piracy update'. Further weapons in Monsanto's armoury to charm and nurture its customers include: a free and anonymous hot line, so farmers can snitch on their neighbours; a posse of Monsanto spies scouring the countryside; and an unpleasant habit of making examples out of growers it catches violating patents.[211] Sometimes that means jail – eight months, plus $1.7 million in damages to Monsanto, in the case of American farmer Kem Ralph[212] – which may strike some people as just a little unjust. In Mexico, Monsanto has really outdone itself – it has taken out ads warning farmers that the growing of its GM crops without contracts could be punished by nine years in jail.[213]

In pursuing a corporate policy (patenting of seeds), which many think is monstrously unfair, not to mention obstructive to progress, Monsanto has managed to make criminals out of innocent farmers, who are just pursuing millennia-old practices like seed-saving. The question is – who are the real criminals here? Who is it that really ought to be fined huge sums of money and thrown in jail?

## The corporate takeover of the food chain, part 2

### Becoming corporate serfs

If this is not warning enough against the further corporatisation of agriculture, GM farmers pay 40–60 per cent more for their seed[214] and are tied into contracts that make them little more than corporate serfs. The Canadian Monsanto agreement states: 'The grower shall purchase and use only Roundup branded herbicide labelled for use on all Roundup Ready canola seed purchased.'[215]

Any farmer who grows GM crops for a year, then reverts to conventional farming, must have his fields inspected by the biotech company for three years to check no GM plants have grown as weeds (volunteers) among the following years' crops – which they almost certainly will have done. In fact, canola seed can remain dormant for ten years.[216] In which case, the farmers are likely to be sued for not paying the royalty. In other words, small farmers cannot revert to conventional farming for fear of being prosecuted. Once you've signed a contract, it's difficult to get out. In fact, often GM and non-GM farmers alike feel the only way of avoiding Monsanto's lawsuits is to grow either more GM crops, rather than less, or simply to abandon non-GM farming altogether.[217]

Percy Schmeiser says this about the Monsanto contract:

To me, it is the most vicious, suppressive contract on the face of the Earth. People don't realize what is going on in North America ... the rights and freedoms of people [farmers] are being taken away. ... The most revolting part of this contract is that you must allow Monsanto's police force to come on your land for three years afterwards, to go into your granaries, with or without your permission to see if you're cheating or not. What I have told farmers ... is to 'never sign that contract, never ever give up the right to use your own seed.' Because if they do, they'll become slaves and serfs of the land.

Percy estimates that at least 40,000 farmers in North America have been investigated, with Monsanto hiring 35 ex-Mounties to do the job – rather too zealously, by all accounts.

Rodney Nelson, an American farmer, said:

We were told that if a farmer represents a field of soybeans to be non-GMO and Monsanto finds as little as *one plant* that tests positive in that field, they may consider that patent infringement... I can also assure you that it is not possible for a farmer in the United States to buy soybean seed from a major seed supplier that is not already contaminated with Monsanto's genetically modified organism (my emphasis).[218]

In other words, Monsanto could successfully sue every non-GM soy farmer in the US.

Monsanto is not alone, though, in this monstrous behaviour, although it has an uncanny knack of being the leader of the pack.

Syngenta has similarly announced that if farmers want its new hybrid barley, they will be obliged to buy Syngenta's chemical products as part of a package. The UK's *Farmers Weekly* condemned the decision, stating that 'it could be the first step towards a future when every input and production method is rigidly controlled from beyond the farm gate. That would reduce farmers to operatives working to the rules of multi-nationals...'[219]

Friedrich Vogel, head of BASF's crop protection business, made it clear that favours to farmers do not form part of the biotech agenda: 'Farmers will be given just enough to keep them interested in growing the crops, but no more. And GM companies and food processors will say very clearly how they want the growers to grow the crops.'[220]

Clearly, the big issue here is the corporate takeover of the food chain. It's one that is suspiciously underplayed by politicians and the media alike.[221] As Remy Trudel, Quebec Agriculture Minister, observed: 'We would be coming back to a situation like the Middle Ages where producers have to depend on a single, powerful company for their livelihood.'[222] Even the biotech companies admit as much:

What you are seeing is not just a consolidation of seed companies, it's really a consolidation of the entire food chain.' – Robert Fraley, Executive Vice-President, Monsanto, 1996[223]

### Making seed-saving illegal

*Iraq:* New legislation in Iraq (Order 81), neatly installed by the Americans, will make the time-honoured pursuit of farmer seed-saving illegal. Yet, an estimated 97 per cent of Iraqi farmers use saved seed. This will create a huge new seed market, effectively handed to the multinationals on a plate. The new legislation conveniently disregards all the contributions Iraqi farmers have made to the evolution of major crops. It also means that they will have to pay for seeds. This will be disastrous for farmers, biodiversity, and the food sovereignty and security of Iraq.

'The US has been imposing patents on life around the world through trade deals. In this case, they invaded the country first, then imposed their patents. This is both immoral and unacceptable,' said Shalini Bhutani, one of the authors of a report by GRAIN and Focus on the Global South.[224] Laureates of the 'alternative Nobel Prize' consider Order 81 a 'crime against humanity'.[225]

GM Watch observes:

The US ... [will] doubtless soon argue, via the World Trade Organization, that the lack of such a law serves as an illegal trade barrier to US companies' patented seeds, and 'harmonisation' will be demanded.[226]

According to a piece in the *Ecologist*:

Under the guise of helping get Iraq back on its feet, the US is setting out to totally re-engineer the country's traditional farming systems into a US-style corporate agribusiness. ... Out will go traditional methods. In will come imported American seeds (more than likely GM, as Texas A&M's [University] Agriculture Program considers itself 'a recognised world leader in using biotechnology' [and is involved with 800 acres of demonstration plots all across Iraq]). And with the new seeds will come new chemicals – pesticides, herbicides, fungicides, all sold to the Iraqis by [US] corporations such as Monsanto, Cargill and Dow.[227]

*Canada:* Canadian farmers are also threatened with losing their traditional right, now conveniently redefined as a 'privilege', to save any kind of seed. There are proposals to collect royalties on virtually all saved seeds, following Monsanto's victory over farmer Percy Schmeiser. The Schmeiser case and the recommendations of the Seed Sector Review are completely contradictory to the International Treaty on Plant Genetic Resources, which came into force in the summer of 2004, says Pat Mooney of the ETC Group, an international research and advocacy organisation. 'The treaty is very strong on farmer seed saving. Canada was the first country to ratify the treaty,' Mooney said. If Canadians allow this to happen, a precedent will have been set for the rest of the world.

'It's a fundamental shift in agriculture to the privatisation of seeds,' says Terry Pugh, Executive Secretary of Canada's National Farmers' Union (NFU). 'There are no benefits for farmers. ... Farmers can't believe this is happening.'[228]

## GMOs IN THE ENVIRONMENT

### Cross breeding with wild relatives

GM contamination not only threatens agricultural biodiversity, but natural biodiversity too. A paper, published in *Science* in October

2003, reveals that GM oilseed rape is prone to widespread cross breeding with wild flowers, notably its relative wild turnip or bargeman's cabbage. An estimated 32,000 plus hybrids would be created along British rivers every year, where it is commonest, and 17,000 in arable areas.[229] These wild flowers might be obliterated by their GM descendants in just a matter of a few generations.

And contrary to the reassuring statements from the biotech lobby, experimental studies confirm that genes passing from crops to weeds can persist for generations, rather than disappearing quickly owing to the lack of any positive selective pressure.[230]

### Disturbing ecological balances

The contamination of complex ecosystems by transgenes is likely to bring unpredictable and uncontrollable knock-on effects. A few of the potential problems follow.

The flow of genes from GM crops to wild relatives – for example, genes giving virus resistance, allowing nitrogen fixation, or giving the ability to grow in marginal conditions – could make them more competitive than, and therefore displace, other wild species. This would reduce biodiversity in a specific ecosystem, and given the interdependence of species, a domino effect could lead to totally unpredictable and dire consequences.[231]

The soil seems to be a potential hotspot for GM problems. A leading DNA expert at Oxford University, Professor Alan Cooper, has warned that it may be too soon to release GM crops and animals, as groundbreaking research has shown that DNA has persisted in some soils for 400,000 years. GM supporters have always claimed that DNA degrades rapidly.[232]

In fact, transgenic DNA has already been found to survive in soil for at least two years, and to replicate during that time.[233] Our knowledge of soil microbiology is not sufficient to predict what will happen when we introduce artificial bacteria into the soil. This seems to be borne out by GM *Klebsiella* soil bacterium – engineered to produce increased ethanol concentrations in fermentors used to create alternative fuel – which is highly competitive and suppresses native soil micro-organisms important for soil fertility. It also causes soil chemistry changes – for example, producing ethanol, a toxin to other micro-organisms and plants.[234] In experiments, it killed wheat plants, when added to the soil, whereas the non-GM *Klebsiella* did not. What would happen if this bacterium was accidentally released? Once established, it could be very hard to eliminate.[235]

Other alarming discoveries in the soil are: the differences found in soil microbial communities around GM canola and conventional canola; and the accumulation of GM Bt toxin in the soil around the roots of Bt plants, where insecticidal activity is retained for at least six months,[236] with possible negative effects on soil fertility.[237]

## Problems with Bt crops

Insect-resistant GM Bt crops contain the Bt gene. But unlike the non-GM version of Bt, which is used as an organic pesticide, the active ingredient in GM Bt crops is switched on all the time. Bt crops have been linked with a number of problems with benign insects, for example:

- A study of lacewings – a predator of crop pests – has shown that Bt maize disrupted development and increased mortality.
- Pollinators like bees might be affected in unpredictable ways by Bt crops; a study showed that bees, fed the active ingredient of an insect-resistant oilseed rape, had problems in learning to differentiate the smells of different flowers.
- Ladybirds fed on aphids, that had been eating insect-resistant potatoes, laid fewer eggs and had half the longevity.[238]
- According to a 1999 published paper from Cornell Professor, John Losey, Monarch butterfly larvae died in significant numbers when they ate milkweed leaves dusted in GM corn pollen (around half the group died within four days, while the rest were half the size of the control group). None of the control group, which were fed non-GM pollen, died.[239]

## Increased herbicide usage

As we have seen, Dr Charles Benbrook's November 2003 report shows that in the US, GM crops have led to the increased use of pesticides. This has resulted mainly from substantial increases in herbicide use on Herbicide Tolerant (HT) crops, especially soybeans, as GM farmers have to spray ever more, as weed populations shift toward tougher-to-control species, and as genetic resistance emerges in some weed species.[240]

This will have resulted in even more chemicals in our food and water and the environment in general. In Britain, the ubiquitous use of agrochemicals in conventional farming has decimated populations of farmland insects (especially bees and butterflies), birds, and wildflowers. And in the US, excessive use of glyphosate (whose use has increased markedly with GM crops), potentially threatens 74 endangered plant species. Fish can be killed by amounts as little as 10 parts per million, and it increases the mortality of earthworms.[241]

## GENETIC POLLUTION IS FOREVER

As we have seen, GM brings with it a mountain of problems; however, it is not sheer volume of problems that is the most alarming bit. Dr Michael Antoniou pinpoints the unique risk of genetic engineering:

Once released into the environment, unlike a BSE epidemic or chemical spill, genetic mistakes cannot be contained, recalled or cleaned up, but will be passed on to all future generations indefinitely.[242]

In other words, it's forever.

Erwin Chargaff, Professor Emeritus of Biochemistry, Columbia University, and the scientist who laid the basis for the uncovering of the double helix of DNA, said:

I have the feeling that science has transgressed a barrier that should have remained inviolate ... you cannot recall a new form of life... It will survive you and your children and your children's children. An irreversible attack on the biosphere is something so unheard of, so unthinkable to previous generations, that I could only wish that mine had not been guilty of it.[243]

I suspect that our grandchildren and great grandchildren are going to look back at this era – if the human race hasn't destroyed itself by then – with a mixture of utter disbelief and savage fury. How could we have invented such a powerful 'technology' and released it with no testing?

# 6

# The Biotech Lobby's
# Dirty Tricks Department, Part 1

Now, the fun starts. This is the bit where you learn how to control the world. Apologies for giving you the abridged version, but a full account would have rivalled *The Lord of the Rings* in length.

## THE CORPORATE TAKEOVER OF GOVERNMENT – MECHANISMS

Really, it all starts with the vast amounts of money that Big Business pours into bribing governments at election time and lobbying governments and regulators in between times. This buys enormous and disproportionate influence. As George Monbiot observes: 'corporate lobbyists [have] infiltrate[d] every particle of government' in the UK.[1] The revolving door phenomenon, where key industry figures are appointed to key government positions, and key political figures are given lucrative industry postings, keeps everyone's bank accounts fat and healthy, and keeps politicians' minds firmly focused on the pro-business track. Influencing the judiciary can get the courts to rule in your favour, as Monsanto found with the US Supreme Court and plant patents. With the EU, the biotech lobby has conducted a 'ruthless ... propaganda campaign';[2] and with international organisations, getting on their committees helps to exert 'undue influence'. Corporate threats can help to influence governments to do their bidding, as Argentina will testify. And undue influence and suspect activities can speed up the commercialisation of GM crops, particularly in the developing countries, where the populations are often not well-informed.

### The US

There is a staggering amount of money swashing around, with the US biotech industry, alone, spending around $250 million a year on promoting GM[3] (that's enough to feed about 3 million malnourished children in Africa every year[4]). This raises the question: if the technology is so good, why spend so much money promoting it?

*Election money*

In the US, from 1989 to July 2003, the biotech companies have donated over $12 million to election campaigns, in individual, PAC (political action committees), and soft money donations. 77 per cent was to Republicans and $3.4 million was during the mid-term 2002 election cycle. Also since 1989, biotech pharmaceutical companies and BIO (the trade group which the industry created in 1993 to represent its interests in Washington) gave a further $13 million in donations. That's a total of $25 million since 1989.[5]

The second highest amount of money given by Monsanto during the 2000 election cycle was received by Larry Combest (R-TX), the powerful Chairman of the House Agriculture Committee. The highest was received by John Ashcroft, (R–MO), the Attorney General,[6] who came up trumps when he asked the US Supreme Courts to uphold the patent rights of the biotech corporations against small farmers.[7]

Tommy Thompson, the Secretary of Health, received $50,000 from biotech companies in election money, and then established a $317 million biotech zone in Wisconsin using state funds[8] – which is a pretty darn good return on your money by anyone's standards.

The twelve members of the Dairy Livestock and Poultry Committee, who between them received $711,000 in PAC money from dairy-related companies, with four of them taking money from Monsanto, stalled a bill sponsored by 181 congressmen in 1994 that required labelling of GM foods. This effectively killed the bill.[9]

*Lobbying money*

Even more money is spent on lobbying. BASF, Bayer, Dow Chemicals, DuPont, Monsanto, and Syngenta spend vast amounts of money influencing regulations that govern their industry. From 1998 to 2002, these companies spent over $53 million lobbying the US federal government – $21 million from Monsanto alone, who was the biggest donor. The rest of the biotech industry spent a further $89 million lobbying Congress, the FDA, and the White House – $14 million from the Biotechnology Industry Organization (BIO), which lobbies and advertises on behalf of the whole industry.[10] That's a total of $142 million between 1998 and 2002.

According to *Capital Eye*, a money-in-politics newsletter published by the Center for Responsive Politics:

In 1999, when the [FDA] began to pay serious attention to the [GM] issue, more than three dozen food and agriculture trade associations formed the Alliance for

Better Foods to promote biotech crops to Congress and the American public. A spokesman for the Grocery Manufacturers of America (GMA) told a Senate panel: 'Acting together, food companies, lawmakers, scientists, farmers and regulators must work to ensure that activists with a political agenda do not kill the promise of biotech foods.'[11]

In 2000, Monsanto spent more than $2 million on lobbying the US government, giving the corporation clear access to officials and negotiators. Monsanto now has members on the Agricultural Policy Advisory Committee for Trade and the official Biotech Advisory Panel. Olivier Hoedeman, from the Corporate Europe Observatory, said: 'The EU and the United States must kick off the habit of listening so closely to large corporations.'[12] Furthermore, in 2002, Monsanto contributed more than $1 million to a campaign to block Oregon's proposed GM food labelling law. And one of Monsanto's former lobbyists, Linda Fisher, was No. 2 at the EPA and a candidate to become the agency's new chief.[13]

And in the case of Bayer, according to CorporateWatch, its clout allows it to

penetrate all major regulatory, standard-setting, legislative, multilateral, and/or governmental institutions. ... [it] can count on the support of ... governments, in particular the US government. Major corporations are often part of the institutions they target, not so much formally but informally (for example, through participation in advisory groups, links with high-positioned bureaucrats and politicians, by drafting proposals and setting agendas, etc.).[14]

*US insider trading/revolving doors*
Personal connections and their role in gaining political support are probably even more crucial than donations and lobbying – which is no doubt why Monsanto maintains 'close ties to policy makers – particularly to trade negotiators'.[15] But how can government decisions possibly be made impartially and objectively when people in positions of high authority frequently have such strong ties to industry?

According to a speech by Dr Ann Clark, of the Department of Plant Agriculture, Ontario Agriculture College:

A brief listing of key people who have migrated between the biotech industry and the USDA, the FDA, or the EPA in recent years would include Linda Fisher (EPA to Monsanto), Val Giddings (USDA to BIO), Terrence Harvey (FDA to Monsanto), Margaret Miller (Monsanto to FDA), Keith Reding (USDA to Monsanto), Michael

Taylor (lawyer representing Monsanto to USDA), and Sally Van Wert (USDA to AgrEvo). These are ... people with key governmental regulatory positions, often over the very products they are/were involved with in industry.[16]

The Bush/Monsanto connection makes particularly fascinating reading. According to Robert Cohen of Pure Food:

Monsanto's lawyer (Clarence Thomas) was appointed to the Supreme Court by George Bush, Sr. The deciding swing vote giving the election to George, Jr was made by Clarence Thomas... Donald Rumsfeld, Secretary of Defense [headed Searle] purchased by Monsanto. ... Ann Veneman, [former] Secretary of Agriculture, was on the board of directors of Calgene Pharmaceuticals, purchased by Monsanto. ... And Mitch Daniels, Director of the Office of Management and Budget ... was the vice president of corporate strategy at Eli Lilly Pharmaceuticals. Eli Lilly and Monsanto developed the genetically engineered bovine growth hormone. Lilly 'owns' the European 'franchise.' Daniels' presence insures that the bovine growth hormone will one day be approved for use in Europe.[17]

As George Monbiot observed: 'Contacts like this have enabled biotechnology companies to achieve an extraordinary degree of access to the US government.'[18]

### The corporate takeover of the judiciary

Influencing the judiciary can be very effective. In December 2001, the US Supreme Court decreed 6 to 2 that any plants may be patented. This effectively gave the green light to GM companies to own the food supply. And it meant that the courts would uphold the patent rights of the biotech corporations against the small farmer. Interestingly, it was Attorney General John Ashcroft who requested the ruling and former Monsanto attorney Clarence Thomas, a Supreme Court judge, who wrote it.[19]

Judge Rodney W. Sippel of the Federal District Court in St Louis (Monsanto's home town) has a record of ruling in favour of Monsanto in various patent cases brought against farmers. The case of Ralph Kem is just one example. Kem was ordered to pay $1.85 million and spend eight months in jail (as a result of two cases). His lawyer said the judge did not allow the defence to present all its evidence, including what it claims is proof that Ralph's signature was forged on the grower licence agreement. Other farmers have made similar claims.[20]

In 1999, a number of America's foremost antitrust lawyers brought a case accusing Monsanto, and other giant seed companies, of attempting to dominate the GM seed market in the 1990s. Judge

Sippel, however, favoured the companies. In September 2003, he decreed that the antitrust case could go ahead, but rejected the section that wanted to compensate growers who lost export markets because of GM crops. The companies had asserted that the seed pricing figures were so diverse and complex that there was no means of proving that large numbers of farmers were afflicted.[21]

A quick look at Sippel's history might explain why he appears to be so sympathetic to Monsanto. Before 1998, when he became a judge, he was employed by the St Louis company now called Husch & Eppenberger. He also worked as an administrator for US Rep. Richard A. Gephardt.[22] Gephardt received $24,675 from Monsanto for the 1994 election and $6,000 in 1996.[23] The law firm of Husch & Eppenberger act for the St Louis corporate elite, including Monsanto. Happily for Monsanto, their licensing agreements give them the choice of having lawsuits heard in St Louis.[24]

## The UK

In the UK, donations to political parties seem to produce some very effective outcomes, and insider trading appears to be somewhat of a national sport in certain circles.

### Donations – Lord Sainsbury

Lord Sainsbury – who has clocked up £11.5 million in donations to Labour over the years[25] – gave the party its biggest-ever single donation in September 1997. The following month, he was made a life peer by Blair and a year later Minister for Science and Technology,[26] where he supervises the allocation of the majority of the government's scientific research budget.[27] Sainsbury is also a member of the cabinet biotechnology committee, Sci-Bio, responsible for national policy on GM crops and foods, and as such is a key adviser to Blair on GM technology.[28]

On being appointed Science Minister, Lord Sainsbury held large shareholdings in two biotech companies, Diatech and Innotech, which were subsequently put in a blind trust. Innotech made him a £20 million paper profit in just four years.[29]

According to GM Watch:

For some, the choice of an unelected biotech investor and food industrialist to be Science Minister, based within the Department of Trade and Industry (DTI), is more than emblematic of the UK's corporate-science culture. ... Mark Seddon, a member of Labour's National Executive Committee, told the BBC, 'In any other country I think a government minister donating such vast amounts of money and effectively buying a political party would be seen for what it is, a form of corruption of the political process.'

Seddon observed that the grassroots thought the party was now 'in the pockets of the powerful and the rich'.[30]

As *The Times* noted in 2002: 'Suspicious minds looked at the 300 per cent increase in the government grant to the Sainsbury Laboratory [of the John Innes Centre, which does research into GM crops] and pondered whether this might be linked to the fact that Lord Sainsbury of Turville is the Science Minister.'[31] The Biotechnology and Biological Sciences Research Council (BBSRC) has also won an extra £50 million in funding since Lord Sainsbury became Science Minister.[32]

Another deeply impressive conflict of interest brought to you by the Labour Party.

### Donations – Lord Drayson

Dr Paul Drayson, head of the BioIndustry Association, knows which side his bread is buttered. He gave a £100,000 donation to Labour. After meeting with Tony Blair in December 2001, the government awarded his company PowderJect a £32 million contract, without any competition.[33]

The entrepreneur's luck continued with his ennoblement to Lord Drayson in July 2004. Six weeks later, the Labour Party received another donation from him – this time for £505,000.[34] And in May 2005, Tony Blair made him a minister at the Ministry of Defence.[35]

### UK insider trading/revolving doors

The Americans are not alone in this phenomenon either – in Britain, the conflicts of interest are equally impressive. Here's a taster from George Monbiot's 16-page 'The fat cats directory':

- Jack Cunningham, MP, Secretary of State for Agriculture and paid adviser to Albright and Wilson (UK) Ltd (an agrochemicals company and member of the Chemical Industries Associate, which lobbies for the deregulation of pesticides).
- Professor Sir John Cadogan, Director-General of the Research Councils (who are supposed to fund scientific work without an obvious or immediate application for corporations) and Research Director of BP.
- Professor Nigel Poole, member of the government's Advisory Committee on Releases to the Environment (ACRE) and External and Regulatory

Affairs Manager of Zeneca Plant Science (which has had six applications to release GMOs approved by ACRE) and sits on five of the taskforces run by EuropaBio (the lobbying organisation that seeks to persuade European governments to deregulate the release of GMOs).

- Professor John Hillman, Director of the government's Scottish Crop Research Institute (which is charged with overseeing government-funded research projects and providing impartial advice on biotechnology to the government) and board member of the BioIndustry Association (whose purpose is 'Encouraging and promoting the biotechnology sector of the UK economy' by lobbying to 'enhance the status of the industry within government').

- Antony Pike, Chairman of the government's Home Grown Cereals Authority (HGCA) (which carries out and funds research into cereal crops. It has so far failed to fund any projects aimed at improving organic cereal production) and Director General of the British Agrochemicals Association Ltd (BAA) and Managing Director of Schering Agrochemicals/ AgrEvo UK Ltd.[36]

## Lobbying the EU

With the biotech industry in deep financial trouble, only a ruthless propaganda machine keeps it alive. EU specialist Steve McGiffen reports on

the most sustained, ruthless and unscrupulous propaganda campaign which even the European Parliament, an institution which works in the face of relentless harassment from corporate lobbyists, has ever witnessed. This campaign, moreover, is not content with spreading lies and confusion amongst legislators in Brussels and other European capitals, it has also kept up a disinformation campaign which has led large numbers of people to believe the exact opposite of a number of clear truths related to GMOs and their dangers.[37]

## The corporate takeover of international bodies

Corporations also put much energy into lobbying international regulatory bodies to remove barriers to corporate globalisation. According to GeneWatch UK, a confidential Monsanto 2000 internal report (about the pursuits of Monsanto's Regulatory Affairs and Scientific Outreach teams for May and June 2000) leaked to them,

reveals that Monsanto is involved in a global campaign to promote GM foods by influencing which experts get on international scientific committees [and] promoting their views through supposedly independent scientists ...

Dr Sue Mayer, GeneWatch UK's Director, said:

The leaked report shows how Monsanto are trying to manipulate the regulation of GM foods across the globe to favour their interests. It seems they are trying to buy influence with key individuals, stack committees with experts who support them, and subvert the scientific agenda around the world.[38]

### Infiltrating the WHO and FAO

In fact, this confidential report describes how Monsanto tried to influence the regulation of GM crops through WHO and FAO committees.[39] An article in the *Guardian* adds that the food and biotech companies have tried to put industry-friendly scientists on WHO and FAO committees. Furthermore, they have financed NGOs who attended formal discussions with UN agencies; they have funded research and policy groups supportive of their opinions; and they have funded individuals who publicly promoted 'anti-regulation ideology', for example in newspaper articles. Thus, just like the tobacco industry did before it, the food industry has succeeded in infiltrating the World Health Organisation (WHO), and exerting 'undue influence' over policies, like reducing fat, sugar, and salt intake levels, which were meant to protect the health of the public.[40]

### Pressuring the WTO

An August 2003 Friends of the Earth International (FOEI) report asserts that Monsanto and the American Farm Bureau Federation (AFBF) – one of Washington's most influential organisations, and one that promotes biotech foods – have been putting pressure on the US administration to employ the WTO to coerce the EU into accepting GM food.[41]

### Lobbying at the Cartagena Protocol on Biosafety

Negotiations of the Cartagena Protocol, held in Montreal in May 1997, were heavily lobbied by 28 agrochemical/biotech companies or company associations, with at least 22 of them from the US and Canada, six from Monsanto alone. And at the January 2000 meeting in Montreal, 31 industry groups were present. But despite this, for the first time ever, the developing countries prevented the developed countries from dominating the negotiation process of an international treaty or agreement. Numerous parts of the protocol are lacking, but its final form offers more good legislation than many believed possible.[42]

## Corporate threats to governments

And if all the more subtle approaches fail, there are always corporate threats to governments.

In 2004, German chemicals company BASF threatened to relocate GM research to other countries if German law continues to restrict R&D into plant biotechnology. In using bully tactics, BASF follows in the footsteps of other GM giants, which have succeeded in forcing agreements with governments on their own terms.[43]

Monsanto has had repeated success in getting Argentinian governments, among others, to do its bidding by such tactics as threatening to withdraw investment and production facilities from the country (see p. 117).[44]

In 1997, Novartis, in a spectacular case of corporate blackmail and arrogance, threatened to withdraw non-GM sugar beet seed from the Republic of Ireland because of its tardiness over GM beet approval, warning that 'Given the importance of Novartis on the Irish market, this would have serious implications for the Irish sugar beet industry [a major crop in Ireland].'[45]

## Corruption and shenanigans in developing countries

Around the world, cases have been recorded of Monsanto or US representatives applying pressure to judicial or national decision-making processes,[46] with Monsanto applying enormous pressure to gain approval for its crops in India and Brazil and eventually succeeding in both.[47] In fact, the leaked Monsanto 2000 confidential report, mentioned earlier, describes how Monsanto tried to influence the regulation of GM crops in 20 countries, for example Thailand, Mexico, Brazil, Korea, Japan, Bulgaria, the US, and the EU.[48] The report also 'reveals that Monsanto is involved in ... gaining influence with key decision makers in government departments in developing countries'.[49]

The US government seems rather partial to using the threat of the WTO, as Croatia[50] and Sri Lanka will testify (see p. 116).[51] The US also seems to be growing fond of using bilateral Free Trade Areas (FTAs) to push biotechnology. There are worries that an FTA signed with Singapore, which demands patents on plants and animals, something that is not even obligatory under the WTO TRIPS agreement, will become a blueprint for the region. Thailand has negotiated an FTA with the US, and the Philippines and Malaysia are in preliminary discussions.[52]

Disasters of all types offer opportunities for corporations to launch seeds and agrochemicals on poor countries, either directly or indirectly through UN agencies like the World Food Programme.[53] Free trial packages may be used to hook farmers initially, along with credit tied to certain products. Slick demonstrations of new technologies convince growers that the corporations have the answers; introducing farmers to herbicides first, prepares the way for GM herbicide-resistant seed in time.[54]

### South Africa

South Africa, which has willingly embraced the austerities of the World Bank et al. in order to get into the Free Trade Club, is the only country in southern Africa prepared to work hand in glove with Monsanto. GM cotton, soy, and maize are now grown in the country, with the first GM crops being grown back in 1998.

According to Mariam Mayet, of the African Centre for Biosafety:

South Africa's Genetically Modified Organisms Act is a poor example of biosafety regulation. It is in effect, merely a permitting system designed to expedite GM imports into the country and releases into the environment.[55]

The government has come under severe attack, from NGO South African Freeze Alliance on GE (SAFeAGE), over its

lack of regulatory transparency, accountability and capacity... The Department of Agriculture's Directorate of Genetic Resources has adamantly refused to share any information concerning existing data on trial and general releases of GE crops. A court case for access to this information is presently under way [May 2003]. The Government is opposing this request and has recently been joined in the case by Monsanto. Moreover the Directorate has regularly admitted that insufficient capacity is in place to inspect, monitor or oversee present GE crops.

SAFeAGE continues:

Extremely limited public input was accepted to guide the laws and regulations that are meant to manage the risks of GE technology; instead they favour those who are introducing GE crops and food into South Africa, [although] public participation in all potentially risky environmental actions is required by law.

The regulations governing how GM crops are grown and overseen are very weak, and 'issues of liability, post harvest follow-up and other implications have been insufficiently addressed'.

Elfrieda Pschorn-Strauss of Biowatch voiced her alarm about how the application for testing of Monsanto's GM cotton, involving multiple 'stacking' genes (that is, containing more than one GM trait), was made known to the public in May 2003: 'The release of [a new and] unknown GMO into the environment is an event of national importance, yet this company is required to place only an advert in a small local newspaper.'[56]

None of this is surprising when you learn who has been involved in setting up the country's regulatory process for GMOs. According to GM Watch:

[Professor] Jennifer Thompson has been ... probably, the key figure in shaping South Africa's regulation of GM crops since its first regulatory body SAGENE, which Thompson chaired, was established under South Africa's apartheid regime. Thompson is still an official advisor on regulatory decisions today. At one and the same time she's a leading figure in a whole series of biotech-industry backed lobby groups, e.g. AfricaBio, ISAAA, the Council for Biotechnology Information and the African Agricultural Technology Foundation.[57]

Then there's Muffy Koch, another lobbyist for the GM industry, who has, in the words of an article in *Australian Biotechnology News,* 'played a key role in developing South Africa's regulatory protocols and legislation governing GM crops'.[58] It's no wonder, then, that the country 'has had such a rapid uptake of GM crops when the line between lobbyist and regulator is non-existent', says GM Watch.[59] South Africa is the 'gateway into Africa', admits Monsanto's MD for Southern Africa, Kobus Lindeque.[60]

Particularly worrying is the bid to extend South Africa's biosafety laws to the whole continent. According to South African environmental and development lawyers, this regime displays a 'cynical disregard' for international environmental doctrines, and the country's development needs.[61]

*Kenya*

Mariam Mayet, of the African Centre for Biosafety, asserts that:

The US Agency for International Development (USAID) [itself funded by biotech companies[62]] is at the forefront of a US marketing campaign to introduce GE food into the developing world. It has made it clear that it sees its role as having to 'integrate biotechnology into local food systems and spread the technology through regions in Africa'. Through USAID, in collaboration with the GE industry and several groups involved in GE research in the developed world, the US government is funding various initiatives aimed at biosafety regulation and decision-making in Africa, which if successful, will put in place weak biosafety regulation and oversight procedures. These biosafety initiatives are designed to harmonise Africa's biosafety laws with those of South Africa's very weak regulations.

Furthermore, she says: 'USAID is also investing heavily in funding various GE research projects in a bid to take control of African agricultural research.'[63] The GM sweet potato project associated with the Kenyan scientist Dr Florence Wambugu is a case in point, with USAID money paying for a three-year postdoctoral position for her with Monsanto, and part-funding the project. The field trials of the project took place in Kenya, and although it was an expensive failure (see pp. 138–9), it has been central to the building up of a biosafety regime in the country. Kenya, South Africa, Nigeria, and Egypt are the only countries in Africa to have formal biosafety regulations, and they were all subjected to strong aid and trade pressure from the US.[64]

The GM sweet potato project (and similarly Syngenta Foundation's showcase Insect Resistant Maize for Africa – IRMA), as Dr Wambugu observed, had many 'spin-offs' for the biotech industry, so that Kenya is now 'well equipped with necessary expertise to serve the needs of [biotech-related] organizations'. For example, the sweet potato project trained many Kenyan scientists and Kenya now has the facilities to run GM field trials and lab facilities where research for other GM crops can take place. Dr Wambugu also asserts that the project has paved the way for more GM crops to be introduced into Kenya, and thence to nearby countries. Although both the sweet potato and IRMA projects are very light on results, they are heavy on PR, and according to Aaron deGrassi, of the UK Institute of Development Studies, this is crucial for an industry 'eager to use philanthropic African projects for public relations purposes. Such public legitimacy may be needed by companies in their attempts to reduce trade restrictions, biosafety controls, and monopoly regulations.'[65]

But just how little popular support this big push for GM has is becoming increasingly apparent. Firstly, a proposal to ban GM crops in Kenya was put forward in 2004 by Davies Nakitare, Saboti Member of Parliament. Secondly, farmers from all over the country were incensed over the 2004 Biosafety Bill to introduce GM crops.

They complained that it allowed no objections nor did it have any provision for compensation in case of problems with GM crops. Others say it ignores the risks of GM foods, particularly those relating to human health and the environment.[66] 'Neither the Kenyan people nor civil society or environmental groups have been consulted in the drafting of the Biosafety Bill,' said Oduor Ong'wen of Southern and East Africa Trade Information Network Initiative (SEATINI). 'Perhaps that is why the Kenyan draft Bill does not even conform to the minimum standards recommended under the international UN Cartagena Protocol on Biosafety, as shown by legal experts affiliated to the African Union.'[67]

### Nigeria

In May 2004, USAID signed a Memorandum of Understanding (MOU) with the Nigerian government, and the Nigeria-based International Institute for Tropical Agriculture (IITA), pledging $2.1 million for agbiotech R&D. According to USAID's mission director in Nigeria, Dawn Liberiover, these funds were to

assist leading Nigerian universities and institutes in the research and development of bio-engineered cowpea and cassava varieties which resist insect and disease pests. ...[and to] improve implementation of biosafety regulations, and enhance public knowledge and acceptance of biotechnology.

In other words, to install weak biosafety rules and win round the population with plenty of spin.[68]

Interestingly, according to GM Watch:

the USAID assistance came shortly after Nigeria's adoption of guidelines on the safe application of biotechnology in the country, and coincided with the opening of discussions between Nigeria and South Africa on the formulation of a model biosafety law, which other African countries can emulate.[69]

However, such pro-GM acts do not meet with public support. In July 2004, the secretary of the All-Nigerian Consumer Movement Union (ANCOMU), Lanre Oginni, urged the government to call off this MOU promoting agbiotech. He described it as ill-considered and said:

There are reasons to be very cautious about genetically modified products as more evidence is emerging that they are not safe. Instead of promoting biotechnology, Nigeria should be investing in sustainable science centres in order to promote and revitalise our indigenous agricultural and health systems in collaboration with the right kind of contemporary western science.[70]

*India*

Monsanto's Bollgard Bt cotton was the first GM crop to be grown commercially in India, in 2002/3. It provides a clear insight into the way the biotech industry behaves in developing countries.

The Bollgard Bt cotton trials, which were undertaken between 1998 and 2002 by Monsanto-Mahyco (Monsanto's Indian subsidiary), were two years shorter than required in other countries. Furthermore, 'the isolation distances maintained were insufficient' and the trials were 'done on very small plots of land' and then extrapolated to commercial scale, according to a 2005 report from the Centre for Sustainable Agriculture (CSA) in Andhra Pradesh. There was also a violation in the law, 'since large scale field trials have to be permitted by GEAC [Genetic Engineering Approval Committee] and not the DBT [Department of Biotechnology]'. The trials 'were not open for independent scrutiny' and the company claimed 'increased yields and cost reductions', but 'the data submitted did not tally with [these] claims'. [71]

So, with the commercialisation of the crop surrounded by so many suspicious circumstances, many questions remain unanswered. Here are a few. Why, for example, was Monsanto-Mahyco entrusted with carrying out its own field trials when India has large, well-funded agricultural research organisations? Why was it allowed to keep the field trials data of Bt cotton secret (which would have exposed the poor performance at an early stage), when there were so many demands to examine it? Why did the government's GEAC release Monsanto's Bt cotton when it was known to be a poor variety and far superior Indian varieties were in the pipeline? Why was approval for commercial cultivation in Andhra Pradesh and Maharashtra granted when no state and district-level committee was set up, as required by the rules under the Environmental Protection Act?[72] And why was the approval in 2002 for commercialisation of three varieties of Bollgard Bt cotton allowed when two representatives of GEAC were absent from the approval meeting? The approval itself was questionable, according to the CSA report, 'since there were legal cases going on against the approval of field trials themselves'. [73]

In August 2003, a leading Indian pro-GM scientist launched a devastating attack on the links between multinational corporations like Monsanto and India's politicians and bureaucrats. Dr Pushpa

Bhargava, described as 'one of the leaders of the biotechnology movement in India', told the hidden story of American efforts to control India's agricultural sector. He talks of inadequate checks and controls on seed companies, of them being allowed to dominate Indian seed business, and of better (and cheaper) local alternatives existing that have been blocked. He states:

We thus never looked at the poor credibility of Monsanto and its widely known and documented habit of misleading and exploiting people and even going against the law. ... there is evidence that Monsanto has falsified its data of trials in India of its Bt cotton. Further contrary to our law the RCGM [Review Committee of Genetic Manipulation] did not make a single site visit during the course of Monsanto's early trials on a limited scale in the country.[74]

In April 2004, P.V. Satheesh of the Deccan Development Society warned that

the powerful industrial lobby in India has been instrumental in a process that might completely dismantle the Genetic Engineering Approval Committee [India's statutory body for GM crop assessment and approvals] ... and hand over the control to an industry-dominated committee in the name of a fast track approval.[75]

Dr Vandana Shiva argues that what is needed is a beefed-up GEAC, not an emasculated one.[76]

The first commercial harvest (2002/3) of Monsanto's GM Bt cotton was a well-documented failure (see Resources, p. 236). The official 2002/3 report of the pro-biotech State Government of Andhra Pradesh (AP) stated that in some areas the net income resulting from Monsanto's GM Bt seeds was between five and seven times less than from indigenous non-Bt varieties;[77] it also mentions that there was a proposal by the State Government to ask Monsanto-Mahyco to compensate farmers for their losses.[78] The 2002/3 report of the Deccan Development Society in AP found that: Bt farmers reported nearly 50 per cent less yield compared to non-Bt farmers; and while only 29 per cent of Bt farmers reported profits, 82 per cent of non-Bt farmers had gained profit.[79] The 2002/3 Greenpeace study of three districts of Karnataka found that for Monsanto's Bt cotton: pesticide and fertiliser costs went up; farmers reported increased labour costs, as the cotton bolls were smaller; seed costs were four times that of non-Bt seed; and market values for Bt cotton were lower than non-Bt cotton hybrids, owing to low quality.[80] Meanwhile, Monsanto-Mahyco produced their own figures, but unfortunately for them, when Greenpeace-India sent its own researchers to check, they found that farmers had been advised by the company to inflate their yield figures, along with many other irregularities.[81]

In its second season (2003/4), Bt cotton again failed economically in Andhra Pradesh, despite the weather being exceptionally favourable for the crop.[82] The AP Coalition in Defense of Diversity (APCIDD) study, by Dr Abdul Qayoom, former Joint Director of Agriculture in AP, shows that Monsanto's Bt cotton was again less profitable than non-GM cotton, by 9 per cent. And again Monsanto presented its own study, which was undertaken by a marketing company who communicated with farmers just once through questionnaires, while the APCIDD study, by contrast, contacted farmers every 15 days. The Monsanto study claimed four times more pesticide reduction than occurred, 12 times more yield, and 100 times more profit.[83]

As a tragic postscript, more than 4,000 farmers have committed suicide in India in the last decade. 'More than 90 per cent of farmers who died in Andhra Pradesh and Vidharbha in the 2005 cotton season had planted Bt cotton', states Dr Vandana Shiva. 'Genetic Engineering is killing Indian farmers.'[84]

### The Philippines

It was only by keeping the public in the dark, and avoiding any debate on the issue at all, that GM corn was approved in the Philippines in December 2002. Fierce opposition included a protracted hunger strike from environmentalists, farmers' groups, and sections of the Catholic Church. Pro-GM pressure came from the prestigious International Rice Research Institute (IRRI), which was developing GM crops.[85] This was helped along by Professor Prakash, who told journalists in Manila in June 2002 that Greenpeace might be funded by companies that would be disadvantaged by GM crops (although he declined to say who these were)[86] and that farmers' post-harvest losses would be stemmed by the longer shelf life of GM crops (a statement that is totally untrue, and not corrected by Prakash if it was incorrectly reported).[87] Furthermore, he made unsupported claims against the critics of GM crops and lauded GM crops as the cure for all the ills of the poor.[88] He also gave evidence to the Philippine Senate Committee on Health.[89]

Allegations of foul play surround the approval process of GM corn in the Philippines. According to Satur Ocampo, of the opposition party Bayan Muna:

The evaluation process was shrouded with charges that the government based its recommendations mainly on data supplied by Monsanto, while totally ignoring the train of scientific evidence showing that Bt corn may trigger a variety of environment food and health hazards.

Marikina Rep. Del de Guzman observed:

Monsanto and BPI [the Bureau of Plant and Industry] reportedly refused to divulge details of the evaluation by consistently invoking so-called confidential business information clauses to keep the inquiring public at bay.[90]

The crop was never safety tested, and even the results of studies testing the efficacy of the crop against a corn pest were withheld from the public. These tests were done in the open air (rather than in glasshouses) and were heavily guarded, in one case by a man accused of infamous human rights abuses under the Marcos regime.

Monsanto's field tests were criticised for contravening local bans for such tests, inflating yield figures, using irrigation to increase yields despite that most Filipino farmers don't have access to this technology, and higher yield figures based on Bt corn being resistant to corn borer, even though the pest is not a major problem in the regions of the Philippines where Monsanto is promoting its GM corn.

Hard-sell tactics were used by Monsanto, including incentives to cooperatives that sell its Bt corn, with one being offered a computer if it sold 100 bags of seed. In one province, the company hired 30–50 sales people, who bore the title 'technician' in order 'to hide behind the credibility of science to sell its dubious products', says Filipino NGO MASIPAG (Farmer–Scientist Partnership for Development, Inc). Furthermore, MASIPAG asserts, in a report entitled 'Selling Food. Health. Hope: The real story behind Monsanto Corporation', that Monsanto has been rather generous to successful Filipino wholesalers, not to mention to scientists, journalists, and religious leaders as well as municipal and provincial officials. The company 'is currently campaigning to receive a 50 per cent [government] subsidy for their seeds'.

MASIPAG further asserts that the revolving door

technique is employed wherever Monsanto goes. In the Philippines, the Deputy Director of ... the Institute of Plant Breeding [IPB] ... is now working with Monsanto, and [another employee] also previously of the IPB has also moved to Monsanto. ... IPB is the country's premier breeding center for crops other than rice. It cooperated with Pioneer and Cargill in field tests for efficacy of Bt-corn, triggering controversy within ... the Filipino scientific community. The project leader for the controversial Bt-corn trials in the Philippines... has also reportedly recently transferred to Monsanto.'[91]

In May 2003, Monsanto, along with the USAID-funded lobby group Agile, was accused of blocking anti-GM bills in Congress by the biggest peasant farmers' group in the Philippines, the KMP. 'Since October 2001, House Bill 3381 ... [which aims to hold off the entry, field-testing, and propagation of GMOs pending a comprehensive safety evaluation] ... House Resolutions 238 and 922 has not moved an inch in Congress because Monsanto and Agile desperately blocks its passage,' said KMP chair Rafael Mariano. 'Even HB 1647 ... which requires the mandatory labelling of all GM crops and GMO-derived food products is being blocked by corporate interest groups and Agile,' said the KMP. Agile has even had an office at the Department of Agriculture, since 1998.[92]

Another lobby group that has been active in the Philippines is the infamous ISAAA. It has been utilising a local farmer to tell the media how wonderful GM crops have been for small farmers, claiming the lobby group's usual massive exaggerations (see p. 24) for the expansion of Bt corn in the country.[93] However, Roberto Verzola explains how that expansion really took place:

The Mon810 Bt corn is distributed in the Philippines without proper labels, under the brand name DK818YG. Farmers are seldom informed that the seeds they are getting are genetically-engineered, or that these are the controversial Bt corn which had been the subject of controversies in the media and public debates. Often, it is the government which buys from Monsanto or its distributors the Bt corn seeds, which are then given for free or at subsidized prices to unsuspecting farmers. Thus, it is only through stealth, deception and the complicity of government technicians, agriculturists and policy-makers that Monsanto has managed to increase the hectarage of Bt corn in the Philippines, at the expense of the Filipino tax payer at that.[94]

*Indonesia*

In a country where state murder and mayhem had been the norm since Suharto took power in the 1960s, it needed the help of the military and a press blackout in February 2001 for Monsanto/Monagro (Monsanto's Indonesia subsidiary) to bring its Bt cotton seed into the country, according to the *Jakarta Post*. The paper reported that

A total of 40 tons of genetically modified Bollgard cotton seed arrived at the Makassar airport from South Africa ... amid strong protests from environmentalists. ... The wide-bodied plane ... was tightly guarded, and reporters and photographers were barred from approaching the plane. Members of the Indonesian Air Force guarding the area said that reporters must back off for security reasons.[95]

Activists were surprised that the seed wasn't quarantined and examined thoroughly before distribution, and they were puzzled by the use of lorries branded 'rice delivery'.

Indonesia was the first South-East Asian country to commercially approve GM Bt cotton, yet the circumstances surrounding its approval were clouded with suspicion. Tejo Wahyu, Executive Director of the National Consortium for the Preservation of Indonesian Forest and Nature, said that by law, to get Ministry of Agriculture approval of the crop, Monagro should have done an environmental impact assessment. 'We are very suspicious why the ministry issued the license (to Monagro),' he said.[96] In September 2001, a coalition of Indonesian environmental NGOs appealed to the Supreme Court to overrule the commercialisation of GM cotton in South Sulawesi, because of the lack of public participation and insufficient environmental impact assessment.[97]

The probable reason for these suspicious events is revealed in a story reported in the *Financial Times* in January 2005:

Monsanto ... is to pay $1.5m in penalties to the US government over a bribe paid in Indonesia in a bid to bypass controls on the screening of new genetically modified cotton crops. ... According to a criminal complaint by the Department of Justice ... the company paid $50,000 to an unnamed senior Indonesian environmental official in 2002, in an unsuccessful bid to amend or repeal the requirement for the environmental impact statement for new crop varieties. The cash payment was delivered by a consultant working for the company's Indonesian affiliate, but was approved by a senior Monsanto official based in the US... The company also admitted that it had paid over $700,000 in bribes to various officials in Indonesia between 1997 and 2002.[98]

As GM Watch observes:

if 'Monsanto finds it necessary to bribe at least 140 officials and family members for half a decade in a country that only makes less than 1 per cent of its overall revenue ... what [can we] expect ... where there's more at stake?[99] ... If they'll go to [these] corrupt lengths ... in search of a regulatory fix, what are the chances that [safety approval] data is not being manipulated when it is totally under Monsanto's control and there are no checks?[100]

Monsanto's GM cotton had trouble living up to the optimistic yield statistics used in its marketing, since in the first planting year, the crop fell victim to drought and a pest infestation. In the second season, Branita Sandhini, a subsidiary of Monsanto's Indonesian subsidiary, doubled the agreed seed price, while at the same time dropping the price it paid farmers for their cotton by 15 per cent. The farmers were unable to refuse these higher prices, since the company was at liberty to refuse to buy their harvest. Ibi Santi, one of the farmers, said:

The company didn't give the farmer any choice, they never intended to improve our well-being, they just put us in a debt circle, took away our independence and made us their slave forever. They try to monopolize everything, the seeds, the fertilizer, the marketing channel and even our life.

Eventually, many farmers refused to pay the outstanding credit, resulting in Monsanto withdrawing from the region.[101]

Then in March 2003, in a last-ditch attempt to salvage its business in Indonesia, Monsanto wrote a letter to Indonesia's Minister of Agriculture, imploring that it was making no money and demanding less regulation. To keep GM crops in play, Monsanto's letter ingeniously proposed that the company supplied 'Bollgard cotton seed to the Ministry of Agriculture, free of royalty fees and at a nominal value for the seed', and that 'the Ministry could offer the seed to farmers through its own distribution network and/or appoint a private partner of assist.'[102] In December 2003, Monsanto abandoned selling GM seeds in Indonesia, saying it could not make a profit.[103]

### Thailand

Thailand has suffered heavy trade pressure from the US since bringing in some GM food labelling and a moratorium on GM crops (January 2002) and crop trials (April 2001). In February 2001, the chief of the Thai Food and Drug Administration, Wichai Chokwiwat, disclosed that trade sanctions had been threatened by a visiting US trade delegation, if the Thais went ahead with labelling, which would affect a massive $8.7 billion a year of Thai exports to the US.[104] In June 2004, the Thai environment minister protested in public against the US insistence that a bilateral free trade agreement was conditional on the country growing GM crops.[105]

Not surprisingly, then, in August 2004, the Thai Prime Minister decided to capitulate to this enormous US pressure and commercialise GM crops, despite massive opposition from farmers, exporters, consumer groups, and environmentalists.[106] Meanwhile, though, contamination of papaya seeds was already occurring as a result of crop trials at a Thai research station.[107] This in turn lead to the delay or rejection of papaya exports to Europe, which cost one company 3 billion baht.[108] Interestingly, the trials themselves were in contravention of an existing ban.[109] The good news is that, catalysed by this potent opposition, ministers refused to go along with the Prime Minister's GM commercialisation proposal, and decided instead to continue the three-year moratorium on GM crop planting.[110]

The biotech corporations appear to have been up to their usual tricks in Thailand, as the following examples indicate.

In common with the Philippine experience, farmers in Thailand (and India) were allegedly given GM seed without being informed of what it was, according to Paul and Steinbrecher.[111] Worse still, in September 1999, BioThai, a Bangkok-based NGO, accused one biotech company of illegally releasing GM cotton, which had not been commercialised. This cotton was found to be growing in fields in central and north-eastern regions of the country. While the company denied this, BioThai stated that 'the evidence clearly points to the company's contempt for Thai laws and sovereignty'. BioThai also asserted that the company had been championing GM crops in the Thai press, using 'advertisements, disguised as newspaper articles. ...they do not mention the various harmful effects associated with GM crops'. [112]

Then, according to the Thai newspaper, *The Nation*, in November 2003:

GM-food producer puts poor farmers in touch with banned technology. Despite the government ban on the production of ... GM ... crops for commercial purposes, [one biotech company] has begun promoting pest-resistant GM corn to farmers in the Northeast. ... 'This is to educate farmers about the technology and let them know that new varieties of corn are available that can solve all their problems,' said the head of the company's government and public affairs division. ... Farmers who participated in the education programme, which was part of the company's marketing roadshow for its seed and agro-chemical products, were given free umbrellas, T-shirts and plastic buckets.[113]

According to the chairman of National Bio Safety Committee (NBC), in September 2004: 'Secret testing of genetically modified (GM) crops prohibited by law are still being conducted under supervision of foreign multinationals, because of loopholes in the law.'[114]

## Pakistan

Pakistan has been so seriously affected by the smuggling of GM seeds that, in September 2002, it decided to legalise GM varieties that had been approved in other countries, in order to keep an eye on what was being imported into the country. Earlier attempts to introduce biosafety regulations had previously stalled.[115]

As far back as August 1999, a government official accused Monsanto of aggressively lobbying the government to bring patenting law into force. The official observed that 'Monsanto is pulling powerful strings to influence the legislative process in its favour, sending letters to government officials, holding meetings with politicians.' Monsanto was worried that loopholes in the TRIPS agreement might favour the rights of farmers over multinationals, since the law would have protected farmers' rights to save, exchange, and reuse seed.[116]

## Sri Lanka

In May 2001, Sri Lanka brought in a ban on GM food imports, in order to make time to study the health risks associated with GM food. The country's Director General of Health Services said at the time that consumer safety was the primary consideration and that the ban would last until GM food concerns were straightened out.[117] Then the USA brought another bit of good ol' American-style democracy to the world by threatening to use the WTO to topple the ban. Under the weight of such pressure, the ban was postponed indefinitely, only two days before its introduction. Local NGO, Environmental Foundation Ltd, tried to get the ban reinstated in early 2002 by lobbying the new Minister of Health; so far, it has been unsuccessful.[118]

## Argentina

Argentina has been a model of compliance with IMF and World Bank regimes for a long time, with an emphasis on commercial-scale export agriculture to service its debt and boost the economy. It has also been the model of obedience to US GM policy.[119] All of which has led it to economic and social disaster, as we have seen.

GM RR soya was first grown in 1996.[120] There was no national debate on GM or any endeavour to inform the public. The approvals of GM crops were granted on the basis of the discredited concept of Substantial Equivalence, and the GM Commission had no NGO representatives on it, while most of its scientists worked for the biotech companies.[121]

Monsanto has been very active in pushing its interests in Argentina, particularly by leaning on the Argentinean government over the last few years.

In December 2000, according to a company spokesperson, Monsanto threatened to shut some of its Argentine enterprises if the government did not lighten regulations on GM food production. According to Reuters:

Argentina's policy of authorizing new GM products only if they have been approved in European Union endangers Monsanto's projects including an $8 million cotton seed processing plant joint venture, said Miguel Potocnik, Monsanto's agriculture director for southern Latin America. 'This investment is in danger and if (the cotton seeds) don't get approved it could be yet another plant that closes in Argentina,' Potocnik told Reuters.[122]

More corporate intimidation followed in October 2003, when Monsanto said it would postpone a $40 million investment because of a 'lack of a clear midterm strategy in the country and lack of adequate intellectual property protection policy'. A local Monsanto official said that for the company to continue investing in the country, it needed 'fair conditions to compete on an equal footing with the other players'. Furthermore, the Argentine seed industry is cooperating with the government to ensure that farmers use certified seed, and to stop the unlawful sale of GM seeds (60 per cent of the RR soy seed market).[123]

In fact, Monsanto claimed in January 2004 that because of a massive black market for GM seeds the company could not make any money. The company said that unless this changes it would refuse to sell new, advanced soy seeds or to study new varieties modified to local conditions. Many suspected that Monsanto was trying to bully the government into controlling farmers more.[124]

In October 2004, Monsanto said it would consider soliciting royalties at ports on soybeans grown in Argentina, something the Agriculture Secretariat regarded as extortion. In fact, the Secretariat was quoted as saying:

Unfortunately, and despite all the meetings and conversations we've had while trying to come up with a legal framework for the sale of seeds, Monsanto persists in its hoodlum-like attitude, one which stands afar from normal business practices.'[125]

Following discussions with the administration, the corporation has agreed to help draw up new laws to control GM seed trade in Argentina.[126]

If this isn't bad enough for the good people of Argentina, a group funded by biotech companies and the wealthiest landlords is giving soy the hard sell, using PR, subsidies, and a media blitz to push the people into consuming more soy products. Since the 2002 economic collapse, the Argentine people have been driven to near starvation, and now they are being force-fed GM soy – designed for cattle not people, and untested for its effects on human health.[127]

*Brazil*

Great pressures have been put on the Brazilian government to commercialise GM, both directly from Monsanto[128] (despite a 1999 court injunction against the release of GMOs)[129] and as a result of the large-scale smuggling of Monsanto's RR soy seeds from adjacent Argentina[130] (which is up to 70 per cent of the crop in the southern state of Rio Grande do Sul).[131]

Despite an election pledge in 2003 to keep Brazil GM-free, President Lula gave in to demands from the US, and the illegal GM soy-growing farmers in the Southern states, and acquiesced to the domestic sale of GM soy for the year 2003–4. Monsanto et al. then jostled for the swift passing of a biosecurity law – one that the Gaia Foundation criticised as 'weak'[132] – and in March 2005, a law was approved, which created the foundation to legalise the sale of GM seeds.[133] Anti-GM organisations said the law was unconstitutional, since there had been no studies on the impact of GM crops on the environment and human health.[134]

## THE CORPORATE TAKEOVER OF GOVERNMENT – RESULTS

As a result of all this money, pressure, and unsavoury activities from the biotech lobby, the industry has realised some major coups around the world. It has managed to avoid proper regulation in many countries and to get loopholes included in those of many more. In the UK and US, government scientific advisory committees are stacked with pro-GM industry figures. In Britain, government funding for non-medical life sciences is skewed in favour of GM and all but ignores organic farming; things are even worse in the US. And the US government seems prepared to go to any length to force-feed the world GMOs, as the last three sections illustrate.

## The US government's weak regulation of GMOs

It's no wonder, then, that the US government deliberately avoided proper regulation of GMOs – a lot of money has been spent seeing to it. And without the compliance of the US government, the business of GM would never have got off the ground. The FDA's false claim about scientific consensus on GM food safety remains the sole purported legal basis for GM foods.[135]

Long before there were any products ready for market, the GMO manufacturers were in Washington taking pre-emptive action to ensure that the regulatory climate would favour their interests. The industry wanted just enough regulation to give the public a sense of assurance, while leaving the manufacturers free of any real restraint. As a result, no new laws were passed governing biotechnology.[136]

Calgene, the manufacturer of the FlavrSavr tomato, was bought by Monsanto in 1997. The company lobbied the US government to weaken the regulatory system, and by February 1992, President Bush Sr declared plans 'to streamline the regulatory process' for bringing GMOs to the market.[137] By a curious coincidence, the FDA, which was in charge of GM food safety, decided, in the same year, that GM food was safe and 'Substantially Equivalent' to non-GM food. As a result, the fox was put in charge of the chicken coup, and the responsibility for GM food safety rested with the manufacturers.[138] Doug Gurian-Sherman, formerly an EPA scientist, observes:

the US Food and Drug Administration does not even approve the safety of GE foods, but instead has a voluntary system of review that is cursory at best, and where the food safety tests are designed and performed by the companies that stand to benefit from the commercialisation of GE crops.[139]

Dr Schubert, of California's Salk Institute for biological studies, mentions that the

data [provided by the biotech crop developer] are not published in journals or subjected to peer review. ... US regulation of GM foods is a rubber-stamp 'approval process' designed to increase public confidence in, but not ensure the safety of, genetically engineered foods.[140]

An article in cropchoice.com adds:

Once crops are released, there is no monitoring or follow-up. Agencies ... ignore significant findings from independent sources, including reports. ...evidence of emerging ... health and ecological problems are routinely disregarded.[141]

In other words, there are *no* mandatory safety testing obligations on the biotech companies in the US. Industry has it all its own way, while government turns a very willing blind eye.

*Steven Druker's evidence*

In 2003, in an explosive press release, US lawyer Steven Druker clearly proves that GM foods have never been adequately tested and regulated.[142] Here are some excerpts:

- GM foods gained approval in the EU only because the US ... FDA ... covered up the extensive warnings of its own experts, blatantly misrepresented the facts, and violated its own laws.[143] ... Nevertheless, FDA bureaucrats, who admit they have been operating under an on-going White House directive 'to foster' the biotech industry ... then declared there is an overwhelming consensus among experts that GM foods are so safe they don't need to be tested, even though they knew their own experts regarded them as uniquely hazardous...[144] And to deepen the deception, they claimed they were not aware of any information showing that GM foods differ from others in any meaningful way. Druker states: 'If the US government had told the truth, no GM foods would have come to market in the US or the EU, since the EU would not have approved them if the US had not.'
- Although the Bush administration attacks the Precautionary Principle as an illegal restraint on trade, it is actually the cornerstone of US food safety law. .... 'Having duped the EU into approving GM foods in the first place, the US is now trying to force open the European market as widely as possible by dismantling the EU's precautionary safeguards and destroying its new labelling laws,' says Druker. 'By allowing them [GM foods] on the market without proof of safety, the FDA is flagrantly breaking it [the Precautionary Principle] in order to promote the interests of Monsanto, Dupont, and other giant corporations.'

EU regulators have themselves joined in the farce, as the following demonstrate:

- 'Unfortunately, EU regulators themselves went on to mislead the public by claiming they're applying the precautionary principle to GM foods when in fact they have been disregarding well-recognized risks and failing to institute the level of safety testing required by their own law,' says Druker.[145]
- Experts with the Public Health Association of Australia (PHAA) thoroughly reviewed many of the data packages the manufacturers submitted to

the regulators and have reported they lack key information ... and that such research could not have qualified for publication in peer-reviewed journals and should not have been accepted by the regulators.[146]

- And a team of Japanese scientists who reviewed Monsanto's tests on its 'Roundup Ready' soybean (which has been approved in the EU) found so many irregularities in the safety assessment they concluded it was 'inadequate and incomplete'. Their November 2000 report concludes: 'The safety assessment of the Monsanto Roundup Ready soybean needs to be reassessed.'[147]

- Moreover, even this deficient data has in many cases revealed potential problems that the regulators have ignored. For instance, the EU authorities have approved several GE foods that are clearly different from their conventional counterparts, even though the differences raise reasonable doubts about safety. These foods include three of Monsanto's GE crops: a variety of maize, a variety of canola, and the Roundup Ready soybean.[148]

### The UK government's GM 'Public Debate'

The Blair government was finally dragged, kicking and screaming, into a 'Public Debate' on GM crop commercialisation, which started in autumn 2002. It consisted of three strands: the Economic Review, the Science Review, and the 'Public Debate'. These were accompanied by the Farm Scale Evaluations (FSEs) – farm-sized trials of GM crops, which ran from 1998 to 2003, and which were the final and most influential part of the investigation into whether to allow GM crop commercialisation. Published shortly afterwards were a number of very damning government reports.

Yet, from the beginning, the whole 'debate' had all the appearances of a sham by a government that seemed set on GM crop commercialisation, said a leading group of independent academic experts – even if, in the end, it failed to hoodwink the British public.[149] Michael Meacher revealed that scientific reports pointing to possible negative effects from GMOs on humans had been 'widely rubbished in government circles'. He also said that the GM food debate had been intentionally stifled with biotech industry pressure. Friends of the Earth's Pete Riley said Meacher's comments established that the GM 'Public Debate' was just a PR exercise to get the thumbs-up for the commercial growing of GM crops in the UK.[150] And to add to all this, as we have already seen, the government's FSA, and the supposedly independent Royal Society, muddied the waters still further with their usual shenanigans (see pp. 40–1, 43–4).

This was an incredibly important debate, given that GM is possibly the most dangerous technology ever invented. So, if the government really was serious about having a proper debate, many questions remain unanswered. For example:

- Why did it allocate so little money to the debate – £200,000 in the beginning, rising to £500,000 under duress[151] – when, in 1999, it was prepared to spend 26 times more on improving the profile of the biotech industry,[152] and when New Zealand (with 14 times less population) spent four times more on a similar process?[153]

- Why did it give so little publicity to the public meetings, give them in such small venues, and with such bad signposting? Public discussions took place in just six towns. The advertising budget was zero.[154]

- Why was the Science Review panel stacked with pro-GM scientists (including scientists from Monsanto and Syngenta), with only two out of 25 critical of GM crops?[155]

- And why, above all, did the government allow a Monsanto director to write the risk assessment chapter of the Science Review?[156] The report concluded there was 'no evidence' GM crops pose a threat to health and the environment, and recommended commercialisation of GM crops on a 'case by case' basis.[157]

- Why did the government allow non-peer-reviewed material as evidence – which is totally unscientific and which the biotech lobby would have screamed bloody murder over if the shoe had been on the other foot?[158]

- Why didn't the government wait for the controversial results of the Farm Scale Evaluations (FSEs) to come in before the 'Public Debate' ended, thus allowing their results to be included?

- And why did the FSEs investigate such a minuscule range of parameters (just the effects of herbicide usage on GM crops on certain wildlife 'indicator' species) and ignore more crucial things like volunteer superweeds, gene stacking, gene flow to soil bacteria, effects on birds, food safety issues, and so on?

- Why, if the government was serious about taking the 'Public Debate' into account, did minister Margaret Beckett, in March

2003, begin the licensing process for 18 applications to grow or import GM crops in Britain, thus pre-empting both the 'Debate' and the FSEs?[159]

- And why even bother commercialising GM when no one wants GM crops anyway?

Even Blair's slick spin machine, spewing black smoke under the strain of it, was unable to put a convincing spin on GM. In the end, the good old British public was not fooled by all the 'weasel words on GM'. As an article in the *Independent* concluded: 'The GM Debate has given the loudest public raspberry conceivable to GM technology.' On a whole series of questions, worded by a government trying to put GM in the best possible light, GM-hostile majorities were enormous, with 86 per cent saying they were unhappy with the idea of eating GM food, 91 per cent saying they thought GM had potential negative effects on the environment, and no fewer than 93 per cent of respondents saying they thought GM technology was driven more by the pursuit of profit than the public interest.[160] Just 2 per cent said they were happy to eat GM food in all circumstances. And the better-informed participants were about GM, the more sceptical they became.[161]

The Economic Review warned that there were no near-term benefits for the UK economy, and that GM crops 'could leave farmers facing a low market price or, in the extreme, no market at all'. It even argues that there could be a 'strong role for the UK to play' in providing non-GM foods.[162]

The Science Review was accused by the Independent Science Panel (ISP), an international group of 24 prominent scientists, of sidestepping the major scientific criticisms of GM – such as its potential health effects and the effects of cross pollination – in its attempt to mislead and cajole the public into accepting the commercial growing of GM crops.[163]

The Farm Scale Evaluations (FSEs) were another complete sham. Yet, despite their ridiculously narrow remit,[164] the results were damning of two of the three GM crops they tested, with Elliot Morley, the UK Environment Minister, saying that results showed 'GM crops had severe implications for wild birds'.[165] Environmentalists warned that the results of the third crop (GM maize) were irrelevant since the non-GM comparison used the highly toxic herbicide, atrazine, now banned in the EU, while the GM crops used a much less toxic herbicide, thus giving the impression of being more environmentally friendly by comparison. Greenpeace's Executive Director Stephen Tindale went further:

The real comparison should be between GM and organic agriculture. But organic is so obviously better for the environment that the GM industry refused point blank to have this included in the trials. The trials are simply comparing one highly damaging form of agriculture with one that's even worse.[166]

Then in October 2003, devastating research published by the UK government showed that pollen from GM oilseed rape travels 26 km, six times further than previously documented, and could contaminate non-GM and organic crops for more than 16 years.[167] In the same month, Tony Blair's chief wildlife advisers, English Nature, warned that the technology would 'seriously degrade' swaths of countryside, and that the use of GM oilseed rape and sugar beet would speed up the (already massive) loss of farmland birds.[168] And in March 2005, Nature reported that the final results of the GM crop trials showed that commercialising some GM crops could change the make-up of weed species in British fields. This resulted in butterfly numbers plummeting by up to two-thirds and bees by a half in crops of GM winter oilseed rape (canola). In turn, this would have 'implications for birds such as sparrows and bullfinches', according to the RSPB's David Gibbons.[169]

Yet despite all this, the Blair government announced on 9 March 2004 that it would proceed with the commercialisation of Bayer's Chardon LL GM maize, without even a debate in Parliament.[170] 'The leaked Cabinet minutes show this to be an entirely political act,' says Geoffrey Lean, an award-winning journalist, 'taken in defiance of the scientific evidence and public concern, by a Government desperate to curry favour with big business, appease President George Bush and, above all, to save the face of a Prime Minister.'[171] And this, we all know, is far more important than protecting the farming industry, the environment, and public health. In the end, there was more than a whiff of similarity between the GM 'Public Debate' and the invasion of Iraq.

But happily, Bayer withdrew its Chardon LL GM maize on 31 March 2004 from the UK (and other European markets), because it had been left 'economically non-viable' owing to the conditions of its limited approval. This means that no GM crops will be grown in the UK until 2008 at the earliest.[172] 'Make no mistake about it, this is a victory for democracy over an arrogant and insensitive biotechnology corporation and over a Government obsessed with a redundant and unwanted technology,' said Dr Brian John of GM-free Cymru.[173] Bayer shares slipped 1.9 per cent with the news.[174]

### UK government bodies bury bad news

Burying bad news is another ploy used by the UK government to hush up unpleasant news. Here are a couple of examples.

DEFRA buried a report on the government's GM crop trials, by publishing only a brief summary on its website on Christmas Eve 2002 (the one day of the year when no newspapers are being prepared). The report was also knowingly held back from the Scottish Parliament and Welsh Assembly until approval of their Draft GM Regulations. Was this because the report indicated that GM rapeseed could not be grown in the UK without massively contaminating non-GM and organic rape? And why did it take so long to publish, given that the trials were finished two years earlier?[175]

In January 2003, the government's Strategy Unit requested comments on the Scoping Note on GM crops, then published them on the internet without stating which organisation the authors were from. Moreover there was no press release either. Was this because the comments were overwhelmingly against GM crops? As Marcus Williamson, Editor of Genetically Modified Food-News, observes: 'In typical UK Government style, the potential release of 'bad news' on GM crops has again been buried.'[176]

### More shenanigans at the USDA

It is no surprise that the USDA is so keen on GM crops, since it will earn about 5 per cent royalties on any GM crops it had a hand in developing.[177] And to keep its employees' focus firmly on the GM ball, they will get a hefty cut too. Craig Sams, Chairman of the Soil Association, and founder of the Whole Earth food company, is not impressed:

the fact that the US government funds research and that USDA employees then get royalties from the patents on that research is an outrage. How on earth can they be expected to make rational choices when a pro-GM choice [which is patentable] can increase their income up to the annual cap of $150,000 a year and a non-GM choice leaves them with nothing but their salary?[178]

## Bush stacks scientific advisory committees

American scientists are growing increasingly worried that the Bush administration is manipulating scientific advisory committees (which should provide agencies with unbiased advice, as well as vetting grant proposals for scientific research) in order to further its political agenda. The Bush administration has put committee members through political litmus tests, eliminating committees whose findings looked likely to disagree with its policies, and stacking committees with individuals who have a vested interest in steering conclusions to benefit industry. 'We've seen a consistent pattern of putting people in who will ensure that the administration hears what it wants to hear,' said Dr David Michaels, a research professor at George Washington University. 'That doesn't help science, and it doesn't help the country.' The Bush administration says it is only doing what every other administration has done in the past, but many scientists take issue with this. 'The Clinton administration did not do this,' said Dr Lynn Goldman, a paediatrician and professor at the Bloomberg School of Public Health at Johns Hopkins University.[179]

## UK government science advisers

The corporate tendrils are well established across the Atlantic too.

Leaked UK government documents showed the extent of the vested interests of supposedly 'independent' government science advisers. Dozens of very influential scientists own shares, enjoy large research grants, or work as consultants for biotech and drug companies. Both former Environment Minister Michael Meacher and Food and Farming Minister Lord Whitty were gravely worried that scientists with commercial links were compromised in giving independent advice, and yet they dominated committees on air quality, food safety, GM crops, and so on. For example:

- Over three-quarters of those on the food safety committee that advises ministers have close ties to large food and drug companies, including Novartis, Astra-Zeneca, and Syngenta.
- A key committee member, who advises ministers on GM safety, was the recipient of research funding from Monsanto and Syngenta. Professor Phil Mullineaux also does GM research at the John Innes Centre, which is financed by Lord Sainsbury.
- Nearly three-quarters of committee members who advise Ministers about the cancer risks of chemicals in food are

employed by large biotech and drug corporations or own shares in them.

Meacher said: 'These committees are absolutely critical. ... I constantly argued that nobody with significant commercial links should be allowed to sit on these bodies. It is vital they are truly independent.' Tony Juniper, Director of Friends of the Earth, said: 'It is now crystal clear how big business is setting the agenda right at the heart of government. ... How can the public trust what Ministers say if their advice is coming from those with vested interest in the biotech or pharmaceutical industry?'[180]

### UK government research funding

As George Monbiot observes: 'The best way of gauging government's intentions is to examine the research it is funding. ... The principal funding body for the [non-medical] life sciences in Britain is the Biotechnology and Biological Sciences Research Council (BBSRC).' And as we have seen, its controlling committees are stuffed with industry executives, and its research funding is heavily biased towards GM and against organic. Monbiot continues: 'What "the market" (which means you and I) wants is very different from what those who seek to control the market want.' The demand for organic food in Britain has never been greater;[181] GM food, by contrast, is about as popular as the plague.

In the US, things are even worse, with less than 0.1 per cent of 30,000 federally financed research projects found to be organic.[182]

### Eat GM or starve

The biotech industry's good friend, the US administration, seems prepared to do anything to foist this unwanted technology on the world.

During the 2002 southern Africa food crisis (which threatened some 13 million) the US forced GM, whole-grain maize aid on recipient countries – even when non-GM was available,[183] and milled grain was acceptable – knowing full well that farmers would save and plant a small portion of seed, thus starting the cycle of contamination. Rafael Mariano, Chairperson of the Filipino peasant farmers' movement, KMP, said: 'The agricultural monopolies are very cruel, knowing that starving people have little choice but to accept the food and be grateful even if our biological future is being slowly corrupted with dangerous technologies.'[184] Greenpeace and ActionAid accused the

US of manipulating the crisis to benefit their GM food interests and of using the UN to dump domestic food surpluses.[185] UK Environment Minister Michael Meacher said: 'it's wicked when there is such an excess of non-GM food aid available, for GM to be forced on countries for reasons of GM politics'.[186] EU Trade Commissioner Pascal Lamy accused the US of being 'very simply immoral' in its linking of the GM issue to food aid, accusing the US of utilising its foreign aid policy to 'dispose of its genetically modified crop surpluses'.[187] Ricardo Navarro, Salvadorian Chairman of Friends of the Earth International, said: 'Food aid is being used, particularly by the US, as a marketing tool to capture new markets. Big agribusinesses are huge beneficiaries of the current food aid system.'[188]

The Bush administration defended itself by going on the attack. It was vociferous in accusing the EU of threatening the lives of millions of Africans by persuading their governments not to accept GM food aid. Nothing was further from the truth – the nations themselves refused it on safety grounds. Michael Meacher accused the US of 'grotesque misrepresentation' in its portrayal of Africa as wanting GM foods. Furthermore, Africa's chief negotiator for the Biosafety Protocol has said that the idea promoted by the US that Africa would accept GM food if Europe did is 'rubbish'.[189]

It was the high moral tone of the US that angered many. If the US really was serious about helping poor nations, why are there so many inconsistencies in US foreign policy? For example:

- Why is the US the stingiest aid donor of any industrial nation?[190] In 2001, the EU provided $23.1 billion in development aid to Africa, compared to just $8.3 billion from the US.[191]
- Why has the US the longest record of taking more money from the developing countries in debt repayment than it gives in aid?[192]
- Why continue subsidising US agriculture when it is one of the main causes of hunger and poverty in Africa? US subsidies encourage surpluses that are then dumped on the world market, depressing world prices and bankrupting African farmers.[193]
- Why is the US promoting Free Trade Areas all over the developing world, when history has shown that protection of infant industries is the only route to achieve strong industry? Is it because free trade usually means a free-for-all for American corporations?[194]

- Why has the US so often sought to undermine the African- and UN-backed Cartagena Biosafety Protocol?[195]
- Why push GMOs at all, if the US is serious about feeding the poor? All the aid agencies and environmental groups, many church groups, and all African countries bar one agree that GM crops will cause more hunger, not less.

And in a postscript that shows the true colours of the Bush administration, in March 2004 USAID stopped all further food aid shipments to Port Sudan, despite being warned by the UN that food stocks for relief operations would be exhausted by April/May 2004. The reason for this is? Because the Sudanese government asked that US commodities be certified GM-free.[196] Similarly, in Angola, the US decreased its food aid after the government banned all GM imports, except for food aid, which it said must be milled (where before it was not).[197]

### Eat GM or die

Probably the most odious ploy of the Bush administration is the US AIDS spending bill of May 2003. The world's press trumpeted it as an act of supreme altruism, while conveniently turning a blind eye to the amendment that ties US AIDS assistance to the acceptance of GMOs. As GM Watch puts it:

In other words, the US is prepared to hold a GM gun not just to the heads of the hungry, but to the desperately ill as well. Either way … the US message is the same: accept our GMOs or we'll leave your people to die.[198]

'Having attempted … to dump unwanted GM maize in Southern Africa [during the 2002 food crisis] they are now resorting to even more unacceptable methods. African nations should have the right to decide what their people are fed. It is immoral for the US to exploit famine and the AIDS crisis in this way,' said Nnimmo Bassey, Director of Environmental Rights Action/Friends of the Earth-Nigeria.[199]

### Eat GM or pay the fine

Unable to get its own way in Europe, the US threw a tantrum and took the EU to the WTO for what it deemed continued restrictions on American GM imports – the de facto EU moratorium on GM food and crops. It seems irrelevant to George Bush that European consumers don't want GM food. Yet the Americans are happy to restrict imports from other countries when it suits them; for example, they have banned cheap Canadian generic drugs because, well … they don't trust them.[200]

If the EU is deemed to be breaking trade rules, the WTO will allow the US to force Europe to approve more GM foods and crops – or face huge financial penalties ($1.8 billion[201]), says Friends of the Earth (FoE) International. FoE Policy and Campaigns Director Liana Stupples said: 'The US Administration, lobbied by the likes of biotech giant Monsanto, is using the undemocratic and secretive WTO to force feed the world genetically modified foods. … Decisions about our food should not be made by the WTO or by Monsanto.'[202] Alexandra Wandel of Friends of the Earth Europe said: 'This trade dispute has Monsanto's finger prints all over it. … Monsanto's influence over past US Administrations is no secret but the impression here is that they have also penetrated the WTO itself. The WTO must purge itself of any such vested interests.'[203]

Happily, though, when the WTO finally ruled, three years later in May 2006, it rejected the majority of US criticisms and failed to advocate action against the EU, stating that under certain circumstances moratoria were permissible.[204] Adrian Bebb of Friends of the Earth Europe said: '… the big corporations that stand behind the WTO failed to get the big win they were hoping for. Free trade proponents needed a clear victory in this dispute to be able to push governments in the EU and the developing world to accept genetically modified food'.[205] John Vidal, of the *Guardian*, thinks: 'It is now clear that the real reason the US took Europe to the WTO court was to make it easier for its companies to prise open regulatory doors in China, India, south-east Asia, Latin America and Africa, where most US exports now go'.[206]

# 7
# The Biotech Lobby's
# Dirty Tricks Department, Part 2

In this chapter, we look at two more important planks in the biotech lobby's ever expanding sphere of influence – the corporate takeover of science and of the media. In the third section, we look at a range of the biotech lobby's dirty tricks, which gives a further taste of how the biotech industry is forcing its products onto the world's unsuspecting population.

## THE CORPORATE TAKEOVER OF SCIENCE

The corporate dollar threatens the integrity of science all over the world. Speaking in the House of Lords, Lord Robert Winston, said: 'above all, we must beware of commercial concerns, which increasingly drive science'.[1]

Corporate control influences what science gets done and what gets reported, confidentiality inhibits the interchange of information, and vicious suppression follows when detrimental results are announced or published, as Professors Pusztai and Chapela can testify (see pp. 142–5). Furthermore, corporate control limits the scope of research to profitable applications (for example, expensive cures to non-lethal ailments of the developed world, while neglecting a vaccine for malaria), many areas of enquiry become off-limits (notably, researching the links between the soaring rates of cancer and industrial pollution), while other contentious key areas are left to amateurs and NGOs (for example, GM pollen flow or the links of child asthma to traffic pollution or studies on corporate crime). As George Monbiot observes: 'Business now stands as a guard dog at the gates of perception. Only the enquiries which suit its needs are allowed to pass.'[2]

Since the 1980s, industry funding for academic research has expanded enormously. In the US, it reached $1.9 billion a year by 1997 – nearly eight times the level of 1977.[3] In the case of the University of California at Berkeley, Novartis came in with $50 million in 1998 and 'bought' the whole college.[4] Nelson Kiang, Professor Emeritus at the Massachusetts Institute of Technology, observes:

You used to have big corporations with labs that would do their own basic research. But ... it's much more effective to turn the universities into R&D labs for them. By sprinkling money around ... they don't have to compete for the best brains in the academic world, they simply buy them at low cost.[5]

As Paul and Steinbrecher comment in *Hungry Corporations*:

TNCs [transnational corporations] can often gain influence over the whole research agenda by merely topping up funds with a small proportion of the total. The universities then provide cheap research and apparently 'independent' advocates for corporate interests.[6]

The same has happened in Britain, with some of the top research universities now depending on private funding for up to 80–90 per cent of their total research budget.[7] George Monbiot notes that 'Britain's universities have swiftly and silently been colonized by corporations.' And 'Today, there is scarcely a science faculty in the United Kingdom whose academic freedom has not been compromised by its funding arrangements.'[8] Gone are the days when public sector research was done for the public good. That fizzled out in the early 1980s when public money for it disappeared and private research began in earnest.

Other problems with the alliance between science and industry are illustrated by various reports and studies:

- The UK Institute of Professionals, Managers, and Civil Servants showed that one-third of government-funded laboratories have been asked to modify their conclusions or advice to: suit the customer's preferred outcome (17 per cent); obtain further contracts (10 per cent); or prevent publication (3 per cent).[9]
- The January 1998 study in the *New England Journal of Medicine* on the impact of conflicts of interest on scientific judgements showed a strong association between authors' published positions on product safety and their financial relationships with the relevant industry.
- More recently, a 2003 study published in the *Journal of the American Medical Association* (JAMA) concludes that industry-sponsored studies are nearly four times more likely to reach pro-industry conclusions than are non industry-sponsored studies.[10]

- In 2002, the *Guardian* revealed that British and American scientists are putting their names to papers they have not written. The papers are 'ghosted' or co-written by employees of the drugs companies, then signed, for a handsome fee, by respectable researchers. In some cases, the researchers have not even seen the raw data on which the papers' conclusions are based. Perhaps 50 per cent of the articles on drugs in the major journals across all areas of medicine are written in this way.[11]
- A study, published in 2001, found that only 16 per cent of scientific journals had a policy on conflicts of interest, and only 0.5 per cent of the papers they published disclosed such conflicts.[12]

Furthermore, massive amounts of UK taxpayers' money have been squandered on research that only benefits commercial interests. We have already seen this in the case of the BBSRC. And between 2000 and 2004, Scottish Enterprise injected nearly $64 million of Scottish taxpayers' money into the development of 'biotech customers'.[13]

Professor Steven Rose, of the Open University Biology Department, said:

the whole climate of open and independent scientific research has disappeared, the old idea that universities were a place of independence has gone. Instead of which one's got secrecy, one's got patents, one's got contracts and one's got shareholders.[14]

Dr David Egilman, a Professor of Medicine at Brown University, Providence, Rhode Island, goes further:

Suppression of science is not an anomaly but is typical of and produced by, the current economic, political, and social situation, and that is – money talks. It is the system; it is not just a few bad apples.[15]

George Monbiot concurs:

The scientific establishment is rotten from top to bottom, riddled with conflicts [of interest] ... There is more corruption in our [UK] university faculties than there is in the building industry.[16]

### The corporate takeover of scientific journals

Again, the corporate dollar poisons all that it touches. Prestigious journals like *Science* and *Nature* are now dependent on corporate advertising and sponsorship, resulting in biased editorial decision-making and falling standards of peer-review. Dr Richard Horton, Editor of the *Lancet*, observed:

Even scientific journals, supposedly the neutral arbiters of quality by virtue of their much-vaunted process of critical peer review, are owned by publishers and scientific societies that derive and demand huge earnings from advertising by drug [and biotech] companies... The pressure on editors to adopt positions that favour these industries is yet another example of the bias that has infiltrated academic exchange. As editor of the *Lancet* I have attended medical conferences at which I have been urged to publish more favourable views of the pharmaceutical industry.[17]

This view seems to be backed up by the following examples of the corporate influence on scientific journals.

*Nature* rejected a peer-reviewed paper confirming the results of Quist and Chapela, on the contamination of native Mexican maize strains by GM maize (see p. 144),[18] and yet it published three articles by Professor Anthony Trewavas – *none based on original research*, but then the biotech brigade never let such trifling details get in their way – promoting GM crops and damning organic agriculture, in a special supplement sponsored by Syngenta.[19]

*Science* published an article by two FDA researchers on Monsanto's GM Bovine Growth Hormone (rBGH) in 1999 that used experimental data from an *unpublished* study *by Monsanto* to show that it had no ill effects on the rats that were fed it, and therefore that it 'presents ... no increased health risk to consumers'.[20] According to George Monbiot, New York's Department of Consumer Affairs review of the paper 'concluded that the data showed "clever deviations from standard and well established safety practice which would be detected only by an experienced scientist"'.[21] In other words, the books were cooked.

*Science* also published Zilberman and Qaim's paper, on 7 February 2003, which referred to the 2001 field trials of Monsanto's Bt cotton in India. Shanthu Shantharam, a strong proponent of GMOs (formerly with the USDA, now working for Syngenta), was surprisingly critical of the piece.

It was ... startling as to how this report passed the muster of peer review at *Science*. This paper really questions the current standards of peer review in a prestigious journal like *Science*... It is unfortunate that both *Science* and *Nature* are making very serious errors in judgement in this most controversial of all technologies... The other weakness of the paper is total reliance on the company (Mahyco) supplied data from field tests and extrapolating [this] into the stratosphere. ... This kind of shoddy publication based on meagre and questionable field data in reputed journals like *Science* does more harm to science and technology development.[22]

In 1999, it was revealed by Dr Richard Horton, editor of the *Lancet*, that a senior Fellow of the Royal Society had threatened him with the loss of his job if he published Dr Pusztai's research on GM potatoes, even though it had been peer-reviewed. The *Guardian* identified who he was.[23]

And when journals, or TV programmes, do put out information that Big Business objects to, the big guns are wheeled out. The case of the *Ecologist* is a classic example. In 1998, the magazine was about to run a special Monsanto edition, when their printers destroyed all 14,000 copies, without warning, for fear of legal action from Monsanto. Zac Goldsmith, the co-editor of the *Ecologist*, observed that the magazine had 'a long history of being forthright about issues and attacking powerful firms, yet not once in twenty-nine years has this printer expressed the slightest qualms about what we were doing'.[24] When the edition finally made it to print, two leading UK newsagents, W.H. Smith and John Menzies, wouldn't sell it for fear also of being sued by Monsanto. Zac Goldsmith commented that

Through reputation alone, Monsanto has been able, time and time again, to bring about what is in effect a de facto censorship. Their size and history of aggression has repeatedly brought an end to what is undeniably a legitimate and very important debate. They believe in information, but only that which ensures a favourable public response to their often dangerous products.[25]

### Bogus science from the biotech lobby

We can see, then, that sound, peer-reviewed, anti-GM science comes under immense pressure from the GM lobby and as a result often doesn't see the light of day. Yet, the GM lobby shamelessly pushes non peer-reviewed, pro-GM rubbish, which erupts all over the media like an unpleasant teenage skin complaint. Here are some examples.

#### Monsanto's Makhathini Flats

In a wonderfully ingenious and cynical PR exercise, Monsanto flew a group of poor South African farmers around the world. They were involved in Monsanto's Makhathini Flats pilot GM cotton project, which started in 1998 and ran for several seasons, with ostensibly small farmers test-growing GM cotton. The farmers were charming the world's unquestioning media with stories of how Bt cotton had upped their productivity and their wealth (well, look, now they could afford flights around the world and fancy hotels), as well as giving a financial boost to their community.[26] This, we were told, was what GM crops could do for the poor of the world.

But this pilot project was guaranteed success from the outset. According to Haidee Swanby of Biowatch South Africa:

Those farmers you see promoting the technology in Europe have had huge assistance from Monsanto in terms of machinery, irrigation, crop management and access to credit. Only this small group will get such assistance which will ensure their yields are good. Others simply find themselves in a debt trap. The wife of the manager of Mboza Clinic, Mrs Nyati, even told me that they were seeing suicides in the area where farmers are drinking their chemicals because of the heavy debt they find themselves in.[27]

A damning report from Aaron deGrassi, a researcher in the Institute of Development Studies at the University of Sussex, states that

These South African farmers ... are plucked from South Africa, wined and dined, and given scripted statements about the benefits of GM. In an area where most farmers cultivate just a few hectares, and only half the population can read, Monsanto's 'representative' farmers are school administrators and agricultural college graduates, owning dozens of hectares of land. Monsanto has been criticized for using these farmers as a part of a deliberate attempt to distort public debate on biotechnology.[28]

Then in June 2005, Biowatch South Africa published a five-year study of GM cotton-growing in Makhathini, which showed that, among many other things, there was no reduction in pesticide usage and that the average debt was $1,322 per farmer, with 80 per cent of them defaulting on their loans. In fact, one farmer commented: 'Four years ago we were told we would make lots of money but we work harder and make nothing.'[29]

And yet, Monsanto – along with biotech industry lobby groups like CropGen, ISAAA, and ABC – have all along shouted long and hard about increased profits, reduction in chemical usage, and environmental benefits.[30] They also just happened to omit any mention of the wholesale interference of an entire floodplain management regime, just to suit the growing of the cotton, with environmental disruption and negative consequences for the farmers downstream.[31]

In other words, this was yet another shameless biotech cooking of the books.

By contrast, Professor Jules Pretty has shown the dramatic results that can be achieved with sustainable agriculture by poor farmers in developing countries without the problems or heavy investments of biotech crops.[32]

### Much ado about a skylark

This massively hyped nonsense really takes the biscuit. The UK's Brooms Barn research station, in research – part-funded by Monsanto,[33] published by the Royal Society, and hawked around the press by Sense About Science (SAS) – found that leaving the weeds uncontrolled between the rows of GM sugar beet until later in the season was good for skylarks.[34] However, there was no data to back this up and the mention of bird life was purely anecdotal; furthermore, the research was based on the fact that sugar beet crops are relatively weed-free, which is simply not the case.[35] Moreover, in the mid-1990s, sugar beet farmers were sold the idea of GM on the basis of its effective weed control, now it was being sold because of its weediness.[36] What they failed to mention was that this weedy regime would be both impractical and undesirable for farmers, since it resulted in a yield loss of up to 31 per cent.[37]

In fact, all this ridiculous cut-it-whichever-way-it-suits hype proves is how damaging conventional chemical farming is, because spraying the rows of crops was found to decrease 'two kinds of insect, and spiders ... sevenfold' compared to the non-sprayed gaps between the rows.[38] Furthermore, this spraying 'transforms fields of crops planted in rows' into 'ecological deserts'.[39] By inference, this also shows how beneficial organic farming is by completely abolishing those agrochemicals – and not just by removing *some* herbicides, from *some* of the field, at *some* periods of the year, as GM *might* do.

The research was roundly slated by anyone who knew anything about the subject,[40] yet once again, with a tedious sense of déjà vu, we discover that it got enormous coverage in the media. This suggests that either the journalists were too lazy to research the veracity of the claims (a quick call to the RSPB or the Soil Association can't be that exhausting, surely?) or, worse still, they reported what Big Business wanted us to hear.

And as if to confirm the thoroughly mendacious nature of this study, later in the year UK government research showed the effects of introducing GM sugar beet could be 'extremely severe, with a rapid decline, and extinction of the skylark within 20 years'.[41]

*More bad science from Brooms Barn*

In more research, this time on the 'Economic consequences for UK farmers of growing GM Herbicide Tolerant sugar beet', Brooms Barn once again excelled themselves, with figures again bearing little relationship to normal farming practice. FARM (a campaigning organisation for British farmers) did an analysis, which discovered glaring weaknesses in the study, for example:

- Herbicide costs for 'conventional practice' have been overstated by 75 per cent.
- Far from reducing costs, the contractual requirements of growing GM HT sugar beet would increase farmers' growing costs by up to £46 per ha.
- The report suggests that Band Spraying could lessen the environmental impact of growing beet, but fails to accurately quantify the impact on yield (acknowledged by Monsanto to be 10 per cent, with other published research suggesting ... up to 31 per cent).
- This suggests a cost of £140 to £430 per hectare in loss of yield alone, quite aside from the increased costs incurred through applying sprays in this way.[42]

*Florence Wambugu's GM sweet potato*

Monsanto's showcase project in Africa – the virus-resistant GM sweet potato – not only wasn't virus-resistant but yielded much less than its non-GM counterpart. Furthermore, the virus it targeted was not a major factor on yield in Africa. You'd never have thought it, though, from the thousands of deferential column inches in the press. These reports suggested huge yield gains, not to mention that this wonder crop would be the saviour of millions of poor Africans. Again, the real facts have been suspiciously ignored by the press.

Dr Florence Wambugu, the Kenyan-born, Monsanto-trained scientist, was able to get remarkable access to American TV, the journal *Nature*, the *New Scientist*, and the *New York Times,* saying things like the biotech revolution could pull 'the African continent out of decades of economic and social despair'.[43] She was even selected as one of 'fifteen people [from around the world] who will reinvent your future', by *Forbes* magazine in December 2002.[44]

Yet, she has been struck down by a curious muteness regarding a poorly resourced Ugandan virus-resistant sweet potato that really is roughly doubling yields, not to mention being overtaken by a galloping senility when it came to the trial stage of her 'high-yielding'

GM crop, which ran from 2001 to 2004, and which in reality yielded less than the non-GM controls.[45] Rather, she claimed that the GM potato could increase yields from 4 to 10 tonnes per hectare,[46] yet forgot to tell us that non-GM potatoes typically yield not 4 but 10 tonnes, according to FAO statistics.[47] Also, there was no peer-reviewed data to back up these claims; as Aaron deGrassi observed in his 2003 report, the researchers simply 'refused to state how the trials, now in their third year, have performed'.[48] Again the Western press lauded the trials, without a shred of evidence, as a wonderful success in feeding the poor of Africa.[49] When the truth of the trials finally came to light it was only with a low-key mention in the Kenyan press and the *New Scientist*.[50]

Monsanto, USAID, the World Bank, and ISAAA have poured an estimated $6 million into this project, which has run for 12 years, and involved 19 researchers, once again diverting both attention and crucial funds from non-GM solutions that would help the billions of poor around the world.[51]

### Doctored science from industry

Probably the most serious case of bad science is the industry's quaint penchant for doctoring studies of GM crops to avoid finding problems with them. Author Jeffrey M. Smith noted the following examples:

- Aventis heated StarLink corn four times longer than standard before testing for intact protein.
- Monsanto fed mature animals diets with only one tenth of their protein derived from GM soy.
- Researchers injected cows with one forty-seventh the amount of GM bovine growth hormone (rBGH) before testing the level of hormone in the milk and pasteurised milk 120 times longer than normal to see if the hormone was destroyed.
- Monsanto used stronger acid and more than 1,250 times the amount of a digestive enzyme recommended by international standards to prove how quickly their protein degraded.
- Cows that got sick were dropped from Monsanto's rBGH studies, while cows that got pregnant before treatment were counted as support that the drug didn't interfere with fertility.
- Differences in composition between Roundup Ready soy and natural soy were omitted from a published paper.
- Antibody reactions by rats fed rBGH were ignored by the FDA.
- Deaths from rats fed the FlavrSavr tomato remain unexplained.

- Aventis substituted protein derived by bacteria instead of testing protein taken from StarLink...[52]

Dr Mae-Wan Ho strongly criticised the data on Bayer's GM Chardon LL T25 maize – used to justify its approval and sponsored by the FSA – revealing that, among other things: a strange study design was used where groups of cows alternated between GM and non-GM diets; GM crops formed only a small proportion of the total diet; and most alarming of all, the T25 maize feeding study was not carried out on T25 maize at all, but on two other GM crops.[53]

In the light of such an abundance of seriously sloppy science, it is hard to take seriously any claims for GMOs by industry-linked scientists.

### Suppressed science – MON863

Monsanto's confidential 1,139-page study on the feeding of GM MON863 maize to rats over 90 days demonstrates that the animals fed on the corn had smaller kidneys[54] and higher white blood cell counts.[55] These problems did not affect the control group that were fed non-GM food.[56]

According to GM-free Cymru, there has been a big cover-up over this report, in order to protect Monsanto and the regulatory bodies that have prematurely declared the maize safe (the European Food Safety Authority (EFSA) okayed it on 19 April 2004[57]). The ESFA, along with the French authorities, would not release the report and the accompanying paperwork, stating that they are covered by the rather convenient 'confidential business interest' (CBI) rules, despite the fact that these rules state that all findings to do with health and safety matters must be made public. The EFSA and other websites contain only information that backs up the EFSA/ACRE position that MON863 is totally safe; all potentially embarrassing information is classified as 'CBI'.[58]

In late 2004, the German government asked Dr Arpad Pusztai to check the report, but before being allowed to see it, he was forced to sign a 'declaration of secrecy' – again, on 'CBI' grounds. Despite Pusztai's serious concerns over the study, the German government would not publish his views and demanded that he abide by his 'gagging order'.[59] But in the end, his comments were made available on the internet. Pusztai said the study had 'many flaws and crucial omissions in it', and 'Overall, this study... has no scientific value.' In discussing the findings, he states:

It is unacceptable ... to regard ... significant increases in white blood cell and lymphocyte counts and decreases in kidney weights in male rats ... as representing normal biological variability ... increased lymphocyte counts are strong indicators of infection or even tumour development. ... the study strongly indicates that feeding rats on diets containing significant amounts of MON863 GM corn ... may cause major lesions in important organs (kidneys, liver, etc), interfere with the function of their immune system (lymphocyte, WBC, granulocyte counts) and change their metabolism (glucose).[60]

In June 2005, a German law court ordered Monsanto to make the report public; the company had tried to block its spread by resorting to the courts. However, an appeal is likely, causing further delay.[61] The same month, the Council of EU ministers decided by a simple (rather than a qualified) majority against the authorisation of MON863,[62] which means that because of the EU's undemocratic and labyrinthine procedures, the unelected members of the European Commission could still yet approve it.[63]

## Anti-scientist smears

One of the key elements of the biotech offensive is to 'shoot the messenger'. If the biotech lobby want to prolong the life of GM 'technology', it is imperative to don the knuckle-dusters and go on the attack against any scientist or campaigner who dares to expose the risks of GM crops. The most high-profile cases of reputations destroyed and jobs lost are those of Dr Arpad Pusztai and Professor Ignacio Chapela. Both discovered devastating problems with GM. Both were on the receiving end of a callous campaign of lies, misinformation, and intimidation. By resorting to technical nitpicking, which the average person couldn't comprehend, and could not therefore challenge, their attackers criticised the scientists for conducting bad science – which is a bit rich given that bad science, as we have already seen, is almost exclusively the domain of the biotech lobby. It is telling that the findings of Pusztai and Chapela are still valid and still part of the scientific body on GM. This proves the mendacity of their attackers, because had their claims been correct, they would certainly have insisted that the science was retracted.

*Dr Arpad Pusztai*

In 1998, 68-year-old Dr Arpad Pusztai, a highly distinguished scientist with more than 270 scientific studies and three books behind him, had been doing ground-breaking research at the Rowett Institute in Aberdeen, which suggested that rats fed on GM potatoes suffered stunted growth and damage to their immune systems.

In August 1998, Dr Pusztai's interview about his research findings went out on the UK TV programme, *World In Action*, and he was congratulated on his performance by his boss Professor Philip James, Director of the Rowett Institute. Yet within 48 hours, Dr Pusztai had been suspended by the Institute and ordered to hand over all his data. He was threatened with legal action if he spoke to anyone, and he discovered that his contract would not be renewed.[64]

The Rowett Institute (dependent on industry funding for its existence[65] and a recipient of a £140,000 grant from Monsanto prior to Pusztai's interview[66]) mounted a character assassination on Pusztai, accusing him, among other things, of mixing up the results and describing him as 'muddled' and 'on the verge of collapse'. None of this was true. Pusztai could not reply to the allegations since he was under a 'gagging order' from the Institute. Yet his reputation was in tatters and he had lost his job.[67] It was only six months later, in February 1999, that 23 scientists from 13 countries formed an independent peer-review panel, which vindicated his work.[68]

What had turned a respected scientist into a pariah overnight? Journalist and author Andy Rowell reports in his book, *Don't Worry, It's Safe to Eat*, and in an article in the *Daily Mail*, that Pusztai's sacking occurred as a result of pressure from the very highest levels of government on both sides of the Atlantic. This in turn came as a result of pressure from a biotech company.[69]

In May 1999, with a precision rarely otherwise seen in Britain, four major reports (the Donaldson/May report; the House of Commons Science and Technology Committee report; the Royal Society review; and the Nuffield Council on Bioethics' report), all congratulatory of GM foods, and all condemning Dr Pusztai's findings, were published within two days of each other. Asked about this, Dr Pusztai commented: 'Can you believe that four major reports could come out, all condemning me, within two days? That is stretching belief.'[70] Minister Jack Cunningham also joined the fray, scorning Dr Pusztai's 'wholly misleading results'.[71]

After such a finely crafted counter-attack, it must have been just a touch embarrassing when, the same week, the British Medical Association (BMA) came out and 'called for a moratorium on planting GM crops commercially' and 'warned that such food and crops might have a cumulative and irreversible effect on the environment and the food chain'.[72]

'Then, as now', Rowell states, 'relationships between senior Labour figures and the GM food companies bordered on the incestuous. In Labour's first two years in office, GM companies met government officials and ministers 81 times';[73] their efforts didn't go unrewarded, because, according to the *Globe and Mail*: 'More than $22 million [was] earmarked in aid for British biotech firms.'[74] Furthermore, as we have seen, Science Minister Lord Sainsbury – an unelected biotech investor and Labour's single largest donor – is one of Blair's key advisers on GM. And David Hill – director of Good Relations, Monsanto's UK PR company – conducted Labour's media campaign for the general elections of 1997 and 2001.

The final twist came when the Pusztais took a holiday to escape for a few days in May 1999. Their house was broken into, but all that was taken was some whisky, some foreign money, and – wait for it – all their research data. Then at the end of the year, the 'burglars' struck at the Rowett Institute, when only the old lab of Dr Pusztai was broken into.[75] You'd have thought that with all those PhDs and all that money they could be a bit less obvious.

After Pusztai's unwelcome discovery about the effects of GM on rats, the government – obviously keen to avoid further embarrassment for industry, and caring not a hoot for public health – ended all public funding of GM safety testing. Interestingly, though, the government was still able to find the money to continue funding the biotech industry.[76]

### Professor Ignacio Chapela

Professor Ignacio Chapela and David Quist uncovered the transgenic contamination of maize landraces growing in remote regions of Mexico.

So serious and contentious was that finding that Chapela went, out of a sense of responsibility, to see a high-ranking member of the Mexican government, before publication of his paper. This was a decision he would soon regret. He was taken to the 12th floor of a deserted building, in the middle of nowhere, where he feared for his life: '[the official] spent an hour railing against me and saying that

I was creating a really serious problem, that I was going to pay for,' Chapela recalled. Chapela then replied: 'So you are going to take a revolver out now and kill me or something...?' In the end, Chapela was offered a cushy position on a scientific team in a secret resort, made up of scientists from biotech corporations, who were 'going to show the world what the reality of GM was all about'. When he declined the offer, Chapela said: '[the official] brings up my family ... [and] makes reference to ... knowing my family and ways in which [the official] can access my family. ...I was scared. I felt threatened for sure.'[77]

Some elements in the Mexican government continued to scare Chapela off publication. Others, however, were worried about his results and so commissioned more research, which backed up his initial findings.

In November 2001, Chapela and his team published in *Nature*. There immediately followed 'every single level of intimidation and aggression that you can imagine', says Chapela. 'It is obviously very well funded and very well coordinated.'[78] Apart from damaging attacks from his own colleagues at the university, some of the worst assaults came from the internet.[79] The attacks on one particular website began within hours of publication, not from a scientist, but apparently from members of the public.[80] These character assassinations catalysed those that followed. Yet, in the end, it turned out that these individuals were actually cyber phantoms invented by a PR company contracted to one of the biotech companies.[81]

*Nature* sort of retracted the paper in April 2002, even though a majority of the journal's own peer-reviewers did not support the action.[82] C.S. Prakash boasted that the AgBioWorld website 'played a fairly important role in putting public pressure on *Nature* ...' and, according to GM Watch: 'has even claimed, in a fund-raising e-mail, that AgBioWorld's campaign led directly to the disavowal of the research'.[83] However, some denounced *Nature* for setting a 'dangerous precedent' and that 'by taking sides in such unambiguous manner, *Nature* risks losing its impartial and professional status'.[84]

In the final analysis, Chapela was denied further tenure by the University of California in December 2003. In other words, he was sacked. Population biologist Professor Wayne Getz, who sat on the committee that recommended giving Chapela tenure, says that Chapela received overwhelming faculty support, but asserts that the review process was 'hijacked' by pro-GM opponents of Chapela at the university.[85] A piece in *Nature* observes that 'Jasper Rine, a geneticist at the university who sat on a key committee reviewing Chapela's tenure, had conflicts of interest.' He 'had financial dealings with biotech firms [and] oversaw the Syngenta agreement', of which Chapela was a strong opponent.[86] In fact, a report, written to evaluate the relationship with the university and Novartis/Syngenta, states that 'there is little doubt' the university's relationship with Novartis influenced the tenure decision.[87] Then finally, in May 2005, justice was done, with the new University of California Berkeley Chancellor, Robert Birgeneau, granting Chapela tenure and retroactive pay.[88]

### Dr Vandana Shiva

Slagging off opponents is all in a morning's work for the biotech boys. Dr Vandana Shiva, the highly acclaimed Indian environmental author, has been slandered as 'villainous' and 'murderous' by biotech proponents, even though she is a world-renowned activist who campaigns for the rights of peasant farmers. Val Giddings, President of the Biotech Industry Organization (BIO), 'honoured' Vandana Shiva with a 'Bullshit Award' made from two varnished piles of cow dung, for her role in 'advancing policies that perpetuate poverty and hunger'.[89] The truth is that once again the GM proponents are projecting their own foul deeds on to their opponents and that the real experts in bullshit are leading GM proponents like Giddings.

### Dr Andrew Stirling

Dr Andrew Stirling was one of only two GM-sceptical panellists on the UK GM 'Public Debate' Science Review Panel in 2003. The other, Professor Carlo Leifert, had already resigned in disgust at the biased and superficial way the review was being conducted and for fear that his research would be jeopardised if he failed to toe the party line.[90] Within days of publication, it became plain that Dr Stirling also felt under pressure. The minutes of a June 2003 meeting revealed how an attempt had been made by a leading pro-GM scientist associated with the review – an individual said to be in a privileged academic and/or regulatory position – to undermine Dr Stirling's reputation and future funding.[91] In other words, Stirling's career was being threatened unless he stopped questioning the technology's safety.

### Professor Trewavas on the attack

In October 2001, the High Court in London named Tony Trewavas, Professor of Applied Biochemistry at the University of Edinburgh, as the origin of a letter containing libellous accusations against Lord

Melchett and Greenpeace, in relation to GM foods and organic agriculture. Trewavas had disseminated written matter slurring their integrity, and has admitted to encouraging others to follow suit, including sending the material to a PR operative and a newspaper editor. A Scottish newspaper, *The Herald*, published this as Trewavas's letter on 3 November 2000; and as late as 6 October 2001, the paper confirmed that it had 'published a letter it had received from Anthony Trewavas'. However, it was only when the newspaper was forced to apologise to Lord Melchett and Greenpeace in the High Court and settle financially that Trewavas finally disclaimed authorship. [92] In correspondence to the *Ecologist*, he said that the letter had been written by a Monsanto PR person.[93]

Trewavas also attacked Professor Chapela, by circulating an email on Prakash's AgBioView canvassing members to write to the University of California to insist on Chapela's sacking if he didn't turn over his Mexican maize samples for independent inspection (that is, by the GM lobby). This is an astonishing attack on the honour of a scientist.[94]

And there was also Dr Pusztai, who, in the same email, was accused of being politically rather than scientifically motivated; Trewavas claimed wrongly that Pusztai had political affiliations.[95]

### THE CORPORATE TAKEOVER OF THE MEDIA

A picture is emerging of large sections of the UK and US media reneging on their responsibilities and not reporting what is really going on – instead, they routinely report the biotech lobby's lies and misinformation and often ignore the surfeit of damning reports and stories on GM. Although the public receive enough generalised information to oppose GM on an instinctive level, not enough details are readily available to avoid it in their food or to know the basic facts – such as, for example, that GM ingredients or material are found in thousands of foods even in the UK, that GM crops do not increase yields, that GM crops do not decrease pesticide use, and that GM will not help feed the hungry. In fact, without the complicity of much of the media (particularly in the US), which has masked the basic failings of GM, Genetic Engineering would barely have got off the starting blocks. The GM debate, therefore, is a superb illustration of how certain sections of the Western media are not objective.

Before looking at specific examples, there follows a quick look at how the truth in general is often edited from our media and why.

Non-news – often in the form of soap-opera style coverage of murders and accidents or the non-stop celebrity psychosis – displaces and takes attention away from real news. For example, we rarely hear about the 34,000 under-5s who die every day around the world from undernourishment, yet if 100 Jumbo jets crashed every day, we'd hear of nothing else.[96]

Omission of epic proportions likewise ensures that we simply don't hear about major global issues. For example, how the developed world is plundering the developing countries. Or how Western-backed dictatorships have committed atrocities across the developing world. Or how, from 1965 to 1980, there was virtually no debate in the mainstream UK media about the nuclear arms race, arguably the most important issue then facing mankind.[97]

The little news there is from most of humanity – Africa represents under 5 per cent of world news[98] – follows a pattern of stereotypes, usually of the-act-of-God or bad-governance variety, which is seriously misleading as to the causes of the vast suffering endured by the large swathes of humankind (Western aid and trade practices immediately spring to mind).

And the routine faithful reporting of industry lies in the press, as we have already seen from many examples in this book, means that industry gets its message heard endlessly and from all quarters, whereas good anti-Big business science is routinely ignored.

The end result is that so often the 'serious' media peddle plausible propaganda and the tabloids sell 'the trifling, the puerile, the trashy and the pornographic', as John Pilger, top journalist, author, and film-maker, puts it.[99] Former US Attorney General Ramsey Clark observes:

The media, owned by the wealthy, speaking for the plutocracy, has the dual role of anaesthetizing the public to prevent serious consideration or debate of such staggering human issues as world hunger, AIDS, regional civil wars, environmental destruction, and social anarchy, and emotionalising the people for aggression, all without a serious military threat in sight.[100]

So, much of the news is tailored to the social, political, and cultural values of media owners – that is, to a pro-big business, right-wing slant. This is because most of the media is owned by fewer and fewer huge corporations. In Britain, Rupert Murdoch controls 34 per cent of the national daily press (the *News of the World*, the *Sun*, and *The Times*), and 37 per cent of the Sunday market;[101] in Australia, he controls seven out of twelve daily newspapers in the various capital cities, and seven of ten Sunday papers.[102] John Malone now owns 23 per cent of all the cable television stations on the planet.[103] Three agencies – Associated Press, Reuters, and Agence France Presse – supply most of the world's 'wire service' news. In television, there are just two agencies providing foreign news footage to all the world's newsrooms – Reuters Television and World Television Network (WTN).[104]

The manipulation of the media is done by varied and plentiful organisations. There is, as we've already seen, corporate propaganda, which rose dramatically in the 1970s and 1980s. Public relations (PR) in Britain and other developed countries has displaced much of the legitimate activity of journalism, with PR-produced material making up half the content of the broadsheets (except in sport) and much higher levels in the rest of press.[105] 'Think tanks', or 'research institutes', mushroomed in the 1970s and 1980s, to support and fund pro-business academics. They are cited all the time and often displace the role of independent journalism.[106] In Britain alone, 90 journalists and broadcasters were disclosed in 1991 as being among the 500 prominent Britons who were/are paid by the CIA.[107] Senior management in the BBC is still vetted by MI5,[108] and from 1945 to 1985 so too were all applicants for editorial jobs.[109]

### Omission by the media

Good old-fashioned omission is the most effective, and probably the least obvious, means of bending the truth, and this is used to particularly good effect in the US. As author Gore Vidal reflects:

The corporate grip on opinion in the United States is one of the wonders of the Western world. No First World country has ever managed to eliminate so entirely from its media all objectivity – much less dissent.[110]

A classic case was the US media coverage of Monsanto's GM Bovine Growth Hormone (rBGH). After Professor Samuel Epstein's July 1989 Op-Ed piece in the *Los Angeles Times*, where he demanded that the hormone was banned 'until all safety questions can be resolved',[111] Monsanto became very proactive, and very successful, in preventing adverse publicity ever happening again. It set up the Dairy Coalition group, which first attacked Professor Epstein's credentials, and then went on to identify and successfully stifle those reporters and reports that were critical of rBGH from getting their message into the mainstream media. As a result, balanced reporting of rBGH was nearly silenced. Another great victory for Monsanto; another sorry tale for democracy.[112]

This lack of coverage continued when GM crops were introduced. The major US media sidestepped virtually any mention of the problems with GMOs until May 1999, almost three years after their commercialisation. Even then, the discussion focused not on human health risks, but on the risk of GM corn pollen to the Monarch butterfly. In fact, incredibly, it was only in 2000, with the huge food recall following the StarLink debacle, that Americans even came to realise they were consuming GM foods on a daily basis.[113] And in April 2002, a study conducted for Food First showed that 'thirteen of the largest newspapers and magazines in the United States have all but shut out criticism of genetically modified (GM) food and crops from their opinion pages'.[114]

Nowhere is this better illustrated than in Steven Druker's May 1998 bombshell news about his lawsuit with the FDA, which showed that their GM policy was against the law. The US media reports focused mainly on side issues, and the *Washington Post*, *New York Times*, and *The Wall Street Journal* did not report it at all. In June 1999, Druker was interviewed by the same three newspapers, but none mentioned the FDA lies and cover-up. It was only three years on, in January 2001, that the *New York Times* finally did a thorough report on Monsanto's undue clout with the FDA. But it was a rare victory against the censorship of the US press. During the four years from 1998, Druker was interviewed and stories were written, only to hit an editorial veto. Anything that made it to print was trivialised and downplayed. There was barely a word on human safety issues by the scientific community. Outside the US, wherever he spoke, the cover-up at the FDA was reported at length.[115] In America, Druker observed: 'it's as if there is an implicit agenda to suppress it'.[116]

## Bad reporting on GM

We have already seen many examples of poor reporting on GM (and other environmental matters), as witnessed by the cases of Florence Wambugu's GM sweet potato project, Monsanto's Makhathini Flats project, Brooms Barn Much ado about a skylark paper, the ISAAA reports, the BBC's *Costing the Earth* and *Counterblast* programmes, Bjorn Lomborg's book *The Sceptical Environmentalist*, and the Johannesburg Earth Summit (see p. 161). But why does this happen? Is it because journalists find it hard to boil down scientific reports accurately? Is it because it is much easier to report myriad pro-GM reports that are sent them by the GM lobby – as witnessed by most of the examples above – rather than researching a story properly? Or is it because reporting a pro-business line is company policy? Dr Tom Wakeford, biologist and action researcher at the Policy Ethics and Life Sciences Research Institute, University of Newcastle, is frustrated by the 'uncritical celebrations of the virtues of GM crops that dominate the media'.[117] Here are a few more of the numerous biotech lies faithfully reported by journalists.

### Repeating Monsanto's 'flawed data'

Following the well-documented disaster of Monsanto's Bt cotton crop in India in 2002/3 – with poor yields, widespread crop failures, many poor Indian farmers facing ruin (see Resources, p. 236) – Monsanto-Mahyco produced its own figures that showed how truly misguided all the other reports had been. And though Monsanto's figures were much criticised for their flawed data and incomplete methodology,[118] they were unquestioningly reported by the Western media. The BBC News reported: 'Cotton crops in India that were genetically modified to resist insects have produced dramatically increased yields.'[119] The *New Scientist* stated: 'Field trials in India suggest that genetically modified crops have far greater benefits in developing countries, than the developed countries for which they were developed.'[120] And '*The Times* has uncritically accepted the findings of the business interest with most to gain from the sale of GM [cotton] technology to India,' according to Charlie Kronick of Greenpeace UK in a letter to the paper. 'To clarify, the paper published in *Science* this month, which your reporter found to be so impressive, is in fact a long distance analysis of trials conducted by Mahyco – a Monsanto subsidiary in India.'[121]

### Mark Henderson, science correspondent at The Times

Mark Henderson, *The Times* science correspondent, has penned a long list of pro-GM articles over the years.

In one of these articles – 'Farmers reap benefit of GM cotton crops' – Henderson gives us a fairytale of happy Indians reaping rich rewards from GM, and yet report after report has shown that, all over southern India, GM cotton has been failing and thousands of farmers have been taking their own lives as a result.[122]

In a 2003 article entitled 'Who cares what "the people" think of GM foods?' Henderson discusses the UK GM 'Public Debate' and states that 'I'm not sure I want the man in the street to set Britain's science, technology and agriculture policy.' He seems to think our views on whether we should have GM crops are a waste of time, and he dismisses anti-GM campaigners, which include numerous scientists, as a bunch of 'Luddites'. Instead, he would rather trust the 'impartial' experts, like the Nuffield Council on Bioethics, who prepared a strongly pro-GM report to coincide with the 'Public Debate'. According to Henderson, this is 'an independent group with genuine expertise'. GM Watch doesn't agree: 'In reality ... the small Working Party of five behind the report is dominated by "experts" with a history of passionate advocacy of GM crops.' Henderson also thinks 'There is nothing good or bad about' GM crops per se, some uses will be good, some bad. Yet, this flies in the face of, and deftly ignores, the conclusions of dozens of independent reports from around the world (see Resources, pp. 234–7).[123]

In a 2002 piece entitled 'Attack on safety of GM crops was unfounded', Henderson wrote that '*Nature* ... admitted ... it had been wrong to publish flawed research', which proved that GM maize had contaminated 'a traditional variety in Mexico'. However, although *Nature* admitted that the paper was deficient in parts, it was not the parts that Henderson referred to. Henderson then wrongly concluded that *Nature*'s judgement of the research represented 'an unprecedented step that weakens the scientific case against the technology'. Yet, the European Environment Agency had just published an important, and thoroughly referenced, paper, which concluded that cross contamination by a number of GM crops, maize included, is virtually inevitable – a fact that Henderson could hardly have been in ignorance of, especially as the research was broadly reported.[124]

In the summer of 1998, in 'Modified crops "help man and wildlife"', Henderson reported at face value Monsanto's alleged environmental benefits for GM sugar beet. With very few questions asked, he reported Monsanto-funded scientists saying that 'Genetically engineered crops can save farmers money, reduce chemical spraying and create a better habitat for birds and insects.' Yet, a few months later in October 1998, the *New Scientist* (and later research) showed these claims to be little more than advertising hype from Monsanto.[125]

In an article from November 2002, 'New GM rice could transform the fight against famine', Henderson reports: 'A strain of genetically modified rice that can grow in droughts, salty soils and cold climates has been developed by scientists, paving the way for a new agricultural revolution for the developing world.' This is a rather confident assertion, given that in his own words: 'No safety trials have yet been undertaken and the plants have not been grown outside a laboratory greenhouse.'[126] Furthermore, even three years later, no GM crop with such capabilities had come on to the market; in fact, GM is proving to be very ineffective at engineering such traits.[127]

### Pallab Ghosh, the BBC's science correspondent

The BBC's science correspondent, Pallab Ghosh, also has quite a track record for pro-GM reporting. Here are some examples:

'The farmers here like genetic modification,' was the opening sentence of a BBC June 2003 report he made about Monsanto's GM cotton in the Indian state of Gujarat.[128] Yet, Ghosh failed to mention that its performance in Gujarat had been so bad that a six-member panel set up by the state government concluded: 'it is unfit for cultivation and should be banned in the State'.[129]

A week earlier, Ghosh had also waxed lyrical about the appalling GM 'Protato' (a 'protein-enhanced' potato), despite the fact that his key claim – about the protato's ability to counter malnutrition – had already been exposed as fraudulent in the Indian press, in March 2003. Ghosh also claimed it was 'expected to be approved in India within six months'. But the Indian press had already stated: 'no request has so far been received from developers for field trials or commercialisation of GM potato and ... it cannot be approved in the current year'.[130] So, why didn't the BBC check out the key claims of a headline UK news story, which was taken up in many other countries?

Ghosh has upset many people with some of his reports:

We are appalled and outraged at the news item 'India's GM Seed Piracy' by Pallab Ghosh, in BBC News on June 17 [2003], which suggests that seed sold by an Indian company Navbharat Seeds, is 'pirated' from Monsanto by farmers of Gujarat. This rumour about piracy is initiated by Monsanto whose Bt cotton has totally failed throughout the length and breath of the country and to divert attention ... Monsanto is trying to focus on the outstanding success [of an indigenously bred cotton variety] ... as unjust and illegal... – Statement with multiple signatories from farmer organisations in India[131]

Ghosh was also behind the BBC's reports that the British Medical Association (BMA) was reviewing its cautious position on GM in January 2003. His claims again hit the headlines, but the BMA issued a press release the same day, which clearly showed the story had not even been checked with them. The BMA labelled parts of his report 'wrong' and 'totally inaccurate'.[132]

Such one-sided reporting is surprising from someone who is the Chairman of the Association of British Science Writers (ABSW). As GM Watch observes:

Ghosh is one of several UK science correspondents whose coverage of the GM issue has led to accusations of bias and an over-cosy relationship with the science establishment and its lobbyists. Others include Mark Henderson at *The Times*, Steve Connor at the *Independent* and Andy Coghlin at *New Scientist*.[133]

### Defamatory claims allowed as Op-Ed pieces

Lord Dick Taverne, the chair of Sense About Science, has run with the biotech brigade smear that anti-GM is akin to terrorism. He wrote in *The Times* that 'Several scientists have received threatening letters, including a bomb threat, for taking a public stand in the GM debate.'[134] According to Jonathan Matthews, founder of GM Watch:

Despite the implication, in this and other articles, of some sort of 'attack' on GM scientists or farmers, no evidence of such a thing has ever been produced. And for a good reason, it has never happened. On the other hand, intimidation of scientists critical of this technology is an issue – an issue that the likes of Taverne are desperately trying to cover up.[135]

The question is – what are the chances of *The Times* allowing a similar Op-Ed piece by anti-GM activists levelling such serious charges at a GM lobby group like Sense About Science? And why is it that the biotech lobby is immune from presenting evidence to back up its case, whereas the anti-GM lobby always has to provide immaculate science before it even stands a chance of being published?

Lord Taverne also penned an article in the *Guardian*, by the title 'The costly fraud that is organic food: its main contribution will be to sustain poverty and malnutrition',[136] which trots out the usual misinformation. No scientific references are given for its claims. Nor is any notice taken of the hard scientific evidence that shows the environmental and health benefits of organic foods.[137] Why did a supposedly liberal paper like the *Guardian* publish a piece like this – and without any evidence?

### Allowing bad news to be buried

If all the omission and bad reporting isn't enough to sound the alarm, there's also the question of why the media's focus seems so often to be where the biotech lobby want it to be. This is especially so when the GM lobby publishes GM events to coincide with reports it wishes to eclipse or ones it wishes to hide behind. For example:

The Royal Society published the Much ado about a skylark piece of junk science (J.D. Pidgeon, 'A novel approach to the use of genetically modified herbicide tolerant crops for environmental benefit', 2002) on precisely the same day (15 January 2003) as a devastatingly critical report on GM was published by the Health Committee of the Scottish Parliament. Was this to divert attention? Certainly the skylark piece seemed to get unprecedented coverage, while the Health Committee report appeared to receive minimal coverage.[138]

In January 2004, the European Commission announced that it had approved Syngenta's GM Bt11 maize for commercialisation – the first new approval since October 1988 – on the very day the much-awaited 'Hutton Report' was published. Was this so the Hutton Report would bury the GM maize scandal? Again the GM maize story was minimally reported.[139]

On 5 March 2004, a paper was rushed online by *Nature* – claiming to show that even with the ban on atrazine, the GM maize would still be marginally better for wildlife – the very same day the Environment Audit Committee (EAC) report was published, which stated the complete opposite. Was this done to upstage the EAC report? The EAC report certainly received much less coverage than it deserved.[140]

### Suspect emphasis and 'independent experts'

We've had omission, bad reporting, and timing, now it's emphasis.

It is interesting how supposedly balanced TV and radio programmes about GM name all the scientific credentials of the pro-GM spokespersons and show them looking important and serious in their labs, whereas the anti-GM voices are often portrayed as woolly and ridiculous. These are regularly shown as worried housewives and members of the knit-your-own-sandals brigade, and rarely the hundreds of highly prestigious anti-GM scientists, who, if mentioned at all, are often stripped of their full titles. References by anti-GM spokespeople to scientific research is often edited out, so that all that's left is the emotive element.[141]

News programmes often report discredited pro-GM propaganda (for example, that GM is needed to feed the hungry) as though it is fact, without a single spokesperson to counter these claims. The public unfortunately trusts the news and if a presenter states something as fact, most people believe it.

Then there's all the 'independent experts', who are wheeled on to TV and radio programmes and used in the press with tedious regularity. In the UK, there are people like Lord May, the President of the pro-GM Royal Society, who is anything but independent.[142] Or the representatives of groups like Sense About Science (SAS), the Science Media Centre (SMC), and the Institute of Ideas (IoI). The media regularly fail to cite the affiliations or funding sources of these providers of information, which helps to increase the illusion that they are neutral or independent, thus removing the necessary context for judging their proffered opinions. Michael Jacobson, CSPI Executive Director, reckons:

If a reporter is going to quote ... non-profit groups funded by corporations, that reporter should be sure to identify the corporations that fund it. ... If a group refuses to disclose its corporate funding, journalists should say so.

CSPI advocates that editors impose strong penalties for failure to divulge conflicts of interest, for example a three-year ban on publication for those authors making incomplete disclosures.[143]

These 'experts' often start the interview stating their independence, then spew out an utterly one-sided and pro-GM line, repeating totally discredited assertions. For example: that 300 million Americans have been eating GM for years, therefore GM must be safe; or that organic farming is dangerous because it uses Bt insecticide as a spray, yet forgetting to tell us that *every* cell of *every* GM Bt plant contains Bt insecticide. This is what Lord May stated with such authority on the BBC's prestigious *Today Programme* in November 2003. Worse still, in the absence of an anti-GM spokesperson, the highly intelligent and well-informed interviewer, Mr John Humphrys, made no attempt whatever to take Lord May to task over his unscientific claims, letting him chunter away to his heart's content.[144] Most alarming of all, this is no isolated incident, either on the *Today Programme* or the UK media in general.

In fact, the *Independent on Sunday* outlined, 'secret meetings in which ministers try to spin the issue [the Pusztai affair], even down to trying to fix which "independent" scientist appeared on the *Today programme* to support the Government line.' The newspaper ended by saying: 'this is the boldest admission so far that [the government] is trying to co-opt [scientists] as part of its PR strategy'.[145]

Clearly, the press needs to assert its independence, or bring the attention of journalists and programme-makers to GM Watch's excellent Biotech Brigade directory (see Resources, p. 237), so they can begin to understand the undisclosed agendas and conflicts of interest of those presented to them as 'independent experts'.

## OTHER DIRTY TRICKS

### Deliberate contamination by the biotech industry

The biotech industry plan has always been to contaminate the world's seed stocks, so that public acceptance of GM becomes irrelevant – the non-GM option will disappear. As Don Westfall, Vice President of Promar, consulting firm to the food industry (clients include Kelloggs, ConAgra, and Aventis), put it: 'The hope of the industry is that over time the market is so flooded [with GM] that there's nothing you can do about it, you just sort of surrender.'[146] Dale Adolphe, boss of the Canadian Seed Growers Association, confirmed this when he said: 'There is so much opposition in the world to any further releases of GM crops that the only way that remains ... is to contaminate.'[147]

This has always been the ploy of the agbiotech empire. The seed companies, now largely owned by the biotech giants, lured us in by saying that GM contamination of seeds wouldn't happen. Now, they say it's inevitable, so we have to accept it. Curiously, though, they seem to control contamination where necessary, for example since EU governments began testing imported seeds, contamination (which had been significant) dropped to under 0.1 per cent.[148]

The US government, always at the service of big business, seems more than happy to help this process of contamination along. The points below certainly suggest that part of its biotech strategy is the contamination of local crop production around the world to the point where governments will no longer be able to ban GMOs.

- One of the most scandalous examples was the admission by the (US-backed) World Food Programme (WFP) that it had been delivering GM food as emergency aid for seven years without informing the recipient countries. Some of this would have been unmilled seed, some of that would have been planted by farmers looking to next year's crop.[149]
- The 2002 drought in southern Africa, as we have seen, is another example of unmilled GM seed being provided as aid. As Ronnie Cummins, National Director of the US Organic Consumers Association, pointed out the US has a record of using food aid to force GM crops 'down peoples' throats'.[150]

- The US also has a record of using exports to force GM crops on unsuspecting nations, as the Mexicans will testify. Scientific analysis shows that one-third of the 5 million tonnes of US maize entering Mexico[151] is contaminated with GM varieties from Monsanto.[152]

Then there are the actions of other governments, and the biotech companies themselves. For example:

- What of the suspiciously minuscule buffer distances for crop trials and commercial crops? Oilseed rape is a classic case. Legal buffer distances were a derisory 5 m in Australia,[153] 200 m in Canada,[154] and 200 m in the UK for trial sites.[155] And yet UK government research has shown that GM rape pollen can travel 26 km.[156]
- Where did the GM seeds come from to plant the 4 million hectares of GM soybeans in Brazil (25 per cent of the crop), given that Brazil had an import and production ban on GM crops from 1998 to 2005?[157] Was it by immaculate conception? Or was it, perhaps, as an 'International People's Tribunal' in the southern Brazilian city of Porto Alegre suggested, from Monsanto and the Federation of Agriculture of the southern Brazilian state of Rio Grande do Sul?[158]
- In Croatia, Pioneer Hi-Bred International was discovered illegally distributing unapproved GM seed to growers and businesses. According to Agriculture Minister Petar Cobankovic, the company will have to pay a fine of about €150,000 and compensate those who purchased its seed.[159]
- And possibly the most pertinent point of all, it has been known for over a decade that GM rape causes widespread and irreversible contamination. This being the case, any growing of this crop must by definition be a deliberate policy of contamination.[160]

### Monsanto's Bovine Growth Hormone (rBGH)

The tale of how Monsanto's GM bovine growth hormone (rBGH) was approved in the US is 'one of the murkiest known episodes in American corporate history', according to George Monbiot.[161]

Bacteria have been genetically engineered to produce large amounts of the hormone rBST (a.k.a. recombinant bovine somatotropin, rBGH, or Posilac). This is injected into cows to make them produce more milk.

However, research on the innumerable side effects of rBGH reads like an A to Z of animal illnesses. They include: making cows more prone to mastitis, and causing foot disorders, reproductive disorders,[162] birth defects,[163] foot and leg injuries, diarrhoea, bloat, indigestion, uterine infections, lesions, metabolic disorder, and shortened lives, plus serious size increases in their ovaries, kidneys, livers, hearts, and adrenal glands.[164] Evidence also shows that rBGH milk may cause breast, colon, prostate, and lung cancer in humans.[165]

The drug is now banned in Europe, NZ, Australia, Japan, and other industrialised countries.[166] However, Monsanto, always in pursuit of its own ends, is attempting to reverse the ban in Europe.[167] The US is the only industrial country where it has been licensed (and it is also sold in a few developing countries).[168] As a result, rBGH is now present in nearly all US dairy products.[169] Sales of the hormone made Monsanto a handsome $200 million in 1998 – a figure that is spookily similar to the amount the hormone costs the US taxpayer every year for the government to purchase excess milk from US farms.[170] Another great value-for-money GM venture for the taxpayer.

In 1985, the US FDA declared that the drug was 'safe for human consumption', even though it had not properly evaluated it.[171] In 1988, Dr Richard Burroughs, of the FDA's Centre for Veterinary Medicine, authorised much more detailed testing of the drug, a decision that led to his dismissal a few weeks later. He later commented that he was sacked, because 'I was slowing down the approval process. ... I don't think the FDA is doing good, honest reviews any more. They've become an extension of the drug industry.'[172]

In 1993, the FDA approved the drug.

However, the entire FDA review of rBGH was described by three members of Congress as 'seemingly ... characterized by misinformation and questionable actions on the part of both FDA and the Monsanto Company officials'.[173] Years later, Canadian scientists put together a long report that detailed the weaknesses, contradictions, gaps, and omissions, in the approval process of the FDA. In essence, the FDA's 1990 evaluation took 'the manufacturer's conclusions at face value'.[174]

Since approval, the FDA has actively promoted the drug,[175] despite plenty of evidence on the health effects on cows and the significant threats to human health. It has also warned shops not to label milk as rBGH-free, so the American people have no way of knowing whether they are drinking rBGH milk or not.[176] And in 1999, two FDA

researchers published a piece in *Science*, saying that rBST 'presents …
no increased health risk to consumers'.[177] (See p. 134.)

In fact, the FDA was perfectly aware of the dangers of rBGH, as
evidenced by stolen documents from the FDA Monsanto files, that
Dr Epstein, Professor Emeritus of Environmental and Occupational
Medicine at the University of Illinois, obtained. These included a
'document from 1987 indicating that the company was fully aware
of rBGH's danger and was conspiring with the FDA to suppress
information critical to veterinary and public health'. Professor Epstein
believes we need Nuremberg-type trials to hold to account industry
executives and scientists, plus everyone else involved in overseeing
public health.[178]

But why is the body charged with protecting the US population
pressurising them to drink potentially hazardous milk? George
Monbiot has the answer: 'The US government is, in cases like this,
simply a channel for corporate power, a vehicle for the global
ambitions of multinational companies.'[179]

On top of all these victories against democracy, Monsanto – never
a company to let trifles like people's health get in its way – filed a
lawsuit against two organic dairies, who had the temerity to label
dairy products as rBGH-free. As a result, both companies went
bankrupt. Monsanto then publicised this to intimidate others from
labelling.[180] The company has also 'threatened school boards with
lawsuits if they ban rBST [milk] from school cafeterias, lobbied against
rBST labelling bills in Congress and states, and threatened states with
lawsuits if they passed rBST labelling laws', according to the Pure
Food Campaign.[181]

Anyone who got in Monsanto's way was bulldozed, as we saw
earlier in the case of the stifling of Professor Samuel Epstein and
any other reporters who wrote about rBGH. There was also the in-
depth TV programme on rBGH that was gagged, as TV reporters Jane
Akre and Steve Wilson, from Fox TV in Tampa, Florida will testify.
Monsanto – one of the TV station's major advertisers – leant heavily
on Fox TV, and Akre and Wilson, to modify the story in some of its
crucial details. In the end, after refusing to be muzzled, Wilson and
Akre were fired by their employers Fox TV, which is owned by another
great friend of democracy – Rupert Murdoch. Akre appealed against
the sacking, but a Florida appeals court ruled against her. The pair
have not only lost their jobs, but they will have to pay Fox's multi-
million dollar legal bill.[182] Fox TV argued that the First Amendment
gives broadcasters the right to even lie or deliberately distort news

reports on the public airwaves.[183] Once again, we see that the US truly is the biggest and best democracy money can buy.

And the UK government – not to be outdone by the shenanigans across the Atlantic – tried a few of its own. Monsanto was granted a secret licence to test rBGH on cattle in Britain, over three years on 38 farms. And although the drug was untested and unapproved, the resulting milk was allowed to be sold on the open market,[184] thus using British consumers as unwitting guinea pigs – which is about all the government seems to think we're good for. And in 1994, the UK was – as always – the sole country that voted against the European Commission moratorium on the drug.[185]

There's much, much more to this story – it's a never-ending tale of dirty tricks, cover-ups, sackings, and misinformation – but sadly I've run out of room.

### Oregon Measure 27

In November 2002, with a record-setting $5–6 million advertising blitz, $1.5 million from Monsanto alone, the biotech lobby defeated Measure 27, the GM labelling proposal for Oregon. In superbly undemocratic fashion, with only a minuscule $5,500 of the funds from Oregon,[186] it turned public support against the measure,[187] despite the fact that almost 60 per cent of Oregonians were in favour beforehand,[188] and that national polls showed 88 per cent of Americans wanted GM foods labelled.[189] The biotech lobby's main weapons were fear and distortion. It claimed that the average food bill would shoot up by $550 a year, whereas independent research showed that labelling would more likely cost up to $10 per person per year.[190]

Paul Holmes, a respected PR business writer, considered the GM industry mistaken in fighting Measure 27, as it would create the impression the industry had a sinister secret to hide. To spend millions of dollars on a campaign to keep their wonderful technology secret, he wrote, is to fly in the face of all logic.[191]

The 'no' campaign had support in the highest places. A letter from the FDA was sent to the Oregon governor, strongly objecting to Measure 27. Lawyer Steven Druker wrote to the governor to warn him that the FDA letter contained 'several major misrepresentations' and falsehoods about GM food safety.[192]

The lack of labelling laws in the US – is staggeringly undemocratic, because it removes the choice not to eat GM foods. As renowned molecular biologist Dr John Fagan said: 'without labelling of GMO products it will be very difficult for scientists to trace the source of new illness caused by genetically engineered foods'.[193] Or as Dr Susan Bardocz, a distinguished senior scientist at the Rowett Institute, said:

They say in the US, everyone has been eating GM food for the past eight years and nothing has happened ... how do they know ... GM food is not labelled. ... That is why industry does not want labelling. They know they can get sued.[194]

Julian Edwards, Director General of Consumers International, makes a searing observation:

One of the ironies of this issue is the contrast between the enthusiasm of food producers to claim that their biologically engineered products are different and unique when they seek to patent them and their similar enthusiasm for claiming that they are just the same as other foods when asked to label them.[195]

**The Fake Parade**

The creativity of the biotech lobby knows no bounds. It will stoop to anything to get us to believe its lies, even using the poorest of the world, willingly or otherwise, as props and stooges to showcase its crops or to imitate popular protest against the anti-GM lobby. This is well illustrated by a couple of gaudy examples that follow.

*The Johannesburg Earth Summit*

At the 2002 Earth Summit in Johannesburg, it wasn't the biggest protest march, involving 20,000 poor, evicted, and landless people that grabbed the headlines around the world, but one that involved a few hundred demonstrators in support of GM crops and free trade. They were supposedly poor and rejecting the 'eco-agenda' of the many NGOs at the summit. Val Giddings, President of the Biotechnology Industry Organization (BIO), seemed to be getting a bit carried away with his own rhetoric when he proclaimed it the 'turning point' for GM. An Indian 'farmer' on the march claimed organic agriculture was causing hunger in India, pleading for GM crops.

The truth was quite different. Yet, once again, the Western press bought into this story without question and without investigation. Here are some facts they didn't mention:

- The figures who organised the rally worked in Washington and London. Their address in Washington was the same as the Competitive Enterprise Institute (CEI) – a powerful lobby group for big business, which receives millions of dollars from corporate America – Dow Chemicals being one of its donors.

- The media contact on the press release, 'Kendra Okonski', is the daughter of a US lumber industrialist. She has worked for a number of far-right NGOs, such as the CEI, and her specialty is helping pro-corporate lobbyists to mimic popular protest.

- The Indian 'farmer' Giddings cites as an example of poor farmers 'speaking for themselves' is Chengal Reddy, who has featured prominently in Monsanto's promotional work in India for at least a decade. He's an affluent right-wing politician who fronts a lobby for big commercial farmers (the Federation of Farmers' Associations – FFA) – and he's never farmed in his life.

- The majority of the 'farmers' were not English speakers, so they were unable to read the anti-environmental messages on the tee-shirts given to them by the organisers. In fact, they weren't farmers at all, but street traders, who thought they were demonstrating for the 'Freedom to trade'. The march organisers' flier omitted to mention the word 'biotechnology'.

- And the poor 'farmers from five continents' – it's hard to imagine how they could afford flights to Johannesburg, on earnings of less than $1/day.[196]

What Jonathon Matthews, anti-GM activist and founder of GM Watch, considers so shameless about these tactics is

the exploitation of situations where it's important – on a life-or-death level – that we are able to discern the truth. It matters what poor farmers and people in the Third World really want, and it matters what actual scientists and real citizens are trying to say...[197]

*New York pro-biotech demonstration*

In late 1999, the *New York Times* reported that a street protest against GM outside an FDA public hearing in Washington was disrupted by a group of African-Americans carrying placards such as 'Biotech saves children's lives' and 'Biotech equals jobs'. However, all was not as it seemed. The *New York Times* learned that Monsanto's PR company, Burston-Marsteller, had paid a Baptist Church from a poor neighbourhood to bus in these 'demonstrators' as part of a wider campaign 'to get groups of church members, union workers and the elderly to speak in favor of genetically engineered foods'.[198]

Apart from the obvious deceit on important issues, it's hard not to ponder how much good these people could achieve if they harnessed this much creativity in more positive areas. I wonder if they sleep well at night?

### Online shenanigans – Monsanto, Prakash, and the Bivings Group

The internet is an important medium for attacks on anti-GM scientists and activists. As Michael Dell, CEO of Dell computers, quaintly observed: 'Think of the internet as a weapon on the table. Either you pick it up or your competitor does – but somebody is going to get killed.'[199]

According to George Monbiot in a 2002 article in the *Guardian*, in the last few years, Big Business has hired PR companies in hundreds of cases to covertly influence the GM debate. Monsanto's online shenanigans included smearing its critics and pursuing a campaign against top scientists, like Professor Chapela, as we have already seen. At the heart of this was Monsanto's Washington-based PR firm, the Bivings Group, with its use 'of websites and bogus citizens' movements which have been coordinating campaigns against environmentalists'.[200] One of Bivings's creations is the Centre for Food and Agricultural Research (CFAR), which is, according to GM Watch, a 'virtual' scientific institute that associates 'Monsanto's critics with violence and terrorism, via lies, innuendo and straight fabrication'.[201]

Monsanto and AgBioWorld were particularly active during the 2002 food aid crisis in southern Africa. In September 2002, for several weeks on the home page of Monsanto-India.com was a link to an article, 'Green killers and pseudo-science', which mixed a description of Johannesburg Earth Summit with a broader assault on 'green fundamentalists' whose 'opposition to genetically modified foods is killing people in famine-hit Africa today, and could threaten Indians in the future too'. The 'green killers' line was one of Prakash's favourites, featuring on AgBioView since the beginning of the trouble.

In late October 2002, the headline Monsanto's electronic newsletter bore was 'Academics say Africans going hungry because of activist scare tactics'. The 'activists' were in fact the employees of a Zambian agricultural college and a Catholic theological establishment. By contrast, the 'academics' included Prakash and Conko, AgBioWorld's founders.

About this time, an AgBioWorld press release insinuated that thousands had died in Orissa, an eastern state of India, because of opposition to GM food aid. In truth, a cyclone had caused the deaths. The email had originated from the IP address of Monsanto Belgium.[202]

### Brainwashing teachers and school children

Brainwashing our children and their teachers, under the guise of 'education', is another oft-used biotech ploy.

In America, since 1991, agribusiness groups have been sponsoring a free, and very comfortable, three-day summer course for teachers in the Chicago area, 'educating' them about the wonders of industrial agriculture, in particular biotechnology and factory farming. All those taking part must agree 'to develop curriculum materials using information obtained from the trip program', which includes brochures showered on them from all manner of agribusiness sources.[203]

Also in the US, in 1999, over 5,000 schools were supplied with a quality magazine, entitled 'Your world – biotechnology and you'. This publication, which lauds GM technology, was created and sponsored by the biotech industry, with the intention of distributing it globally.[204] In Scotland, where over 140,000 copies of the publication were shoved into Scottish schools by Scottish Enterprise in 2001, it has drawn much protest from teachers, environmentalists, and consumer groups.[205]

In Britain, the John Innes Centre (JIC), which receives funding from just about every major biotech company, piloted 'Biotechnology in our food chain', an 'online information service', at two schools in the Norwich area in Spring 1998. It was largely funded by Lord Sainsbury's Gatsby Trust (which puts millions into the study of plant genetics). Critics were less than convinced that it gave a balanced look at the subject. The JIC also commissioned a play, called 'Sweet as you are', to sell GM in UK secondary schools in Spring 2000. Dr Jeremy Bartlett, a scientist in plant genetics from John Innes, attended the play and observed that:[206] 'The GM campaigner looks ridiculous ... and his fiancée listens to the rational scientist and furthers her career by promoting GM foods.'[207]

# 8
# Setbacks for the Biotech Lobby

Despite the wealth and influence of the biotech giants, and the massive corruption of the political process that this has bought, particularly in the US and Britain, the scale of popular protest around the world has slowed the invasion of GM crops and food enormously. Monsanto's ideal world view was that in fifteen to twenty years 100 per cent of commercial seeds would be GM.[1]

Writing ten years on from the first commercialisation of GM crops (1996), things look rather different. GM crops are grown on a mere 5 per cent of the world's cultivable lands,[2] and only on a meaningful scale in six countries. Most of these, bar the US, are having serious doubts about the technology. Furthermore, there are only four major GM crops, and these are used primarily for animal feed or fibre.

In addition, as we have seen throughout the book, new crop candidates for genetic engineering are dropping like flies. Here, by way of a happy round-up, are a few examples:

- A few years ago, Monsanto dropped its GM potatoes after US fast-food companies said they would not buy them.[3]
- In 2003, Monsanto dropped plans to use GM crops to produce pharmaceuticals.[4]
- In March 2004, Bayer withdrew from growing GM Chardon LL T25 maize in the UK and other European markets.[5]
- In April 2004, Spain withdrew Syngenta's Bt176 corn from the market, after a request from the EU over concerns that the crop might cause resistance to antibiotics.[6]
- In May 2004, Syngenta's GM maize Bt11 was approved in the EU, but was later voluntarily withdrawn by the producer following strong consumer resistance.[7]
- In May 2004, Monsanto was forced to shut down its programme to introduce GM canola to Australia. This was just four months after it was commercialised by the federal government – the decision was overruled by every state that could grow the crop, with bans or moratoria.[8]

- Also in May 2004, Monsanto withdrew from GM wheat worldwide. This was a particularly important and significant victory, because it was the first major food crop up for commercialisation and the whole food industry was united against it. This was in part because the technology is clearly not working – after eight years of commercial GM crops, and a barrage of damning reports, farmers didn't want to go near it – and in part because of massive consumer opposition to GM foods.

In Britain – again thanks to massive public resistance – not a single GM crop has yet been grown commercially. At the end of 1996, it was thought to be barely a year away from widespread cultivation of GM rape across the UK, but now it is unlikely there will be commercial GM crops grown before 2008 at the earliest (which is also true across the EU). Furthermore, in both 2004 and 2005, there was not a single field trial registered in Britain,[9] and in 2005, there ceased to be any GM varieties left in the seed-listing pipeline.[10] In fact, by 2004, Syngenta, Monsanto, Bayer, and DuPont had all deserted their GM activities in Britain,[11] with all industry-funded research having closed down,[12] leaving only academic centres doing this work.

Here, to warm the cockles of the heart, is a brief and very partial summary of the victories of ordinary people against some the most powerful corporations and governments in the world.

### GM-FREE AREAS

In response to public consternation at the rapid deployment of GM crops, GM-free areas are springing up all over the place.

According to a 2004 article in the *Guardian*, in Europe,

more than 1,000 French town mayors ... half of all Greek prefectures and nine out of 10 regions of Austria that are all requesting [GM] bans in their areas. In Britain, 12 county councils, nine unitary authorities, two metropolitan districts, one London borough, 13 district councils, two national park authorities, and 35 Welsh councils have voted against the crops. That means that about 14 million people in Britain are living in areas with a GM-free policy.[13]

Margaret Beckett, the Environment Secretary, has conceded that the UK government may have to allow GM-free zones because of public opposition.[14]

In June 2003, a GM-free organic region was declared in all of Slovenia, and parts of Austria and Italy.[15] By 2005, in Europe as a whole, there were 172 regions/provinces[16] and 4,500 sub-regional areas that had declared themselves GM-free,[17] with 70 per cent of Italian regions, 27 of its provinces, and 2,000 towns and cities following suit.[18] Other GM-free areas include: parts of Germany, Switzerland,[19] Tuscany, and the Basque Country,[20] plus Brazil's number two soy-growing state Parana,[21] Bohol province in the Philippines,[22] Powell River GM-free crop zone in Canada,[23] Bolivia, and Croatia.[24] The Swiss are to have a referendum, which, if successful, will ban GM products from Switzerland for five years.[25] Zambia intends to become GM-free.[26] And, in April 2005, 1,000 GM-free zones were declared in Ireland.[27]

China is aiming at having the world's largest area of non-GM soybeans by 2008, which will replace more than 10 million tons of imports a year (much from the US) and make the country self-sufficient.[28]

## ATTACKS ON FIELD TRIALS

As governments ignore what their citizens want so attacks on field trials increase.

In France, in the summer of 2003, the Confédération Paysanne intensified its anti-GM actions, in view of the imminent lifting of the EU de facto moratorium on GM crops. 'We want to show that our determination is as great as ever ... in the battle against GM,' says Bruno Galloo, head of the Confédération Paysanne for Picardie. Part of the strategy was to alert the public to the dangers of GM pollen spread. Bayer said in a statement that it was beginning to question whether it was worth continuing GM trials on French soil.[29]

In Britain, in September 2003, Bayer announced it was pulling out of GM crop trials in Britain, at least until conditions were 'more favourable', because 'they are always ripped up'. This is more evidence that direct action is effective.[30]

In India, in September 2003, more than 40 Indian farmers destroyed a former Monsanto research centre in Bangalore. The company said: 'One greenhouse was destroyed. We have lost valuable plants ... [and] are seriously concerned.'[31] The protestors were enraged because more than 70 farmers in the region had committed suicide in just three months. This was blamed on drought, debt, and GM crops from Monsanto. Professor M.D. Nanjundaswamy, head of the

Karnataka State Farmers Association, said the protestors were sending a warning to Monsanto to leave the country. Several farms trialling Monsanto's GM cotton crop had also been torched by the farmer's association.[32]

In Brazil, in June 2003, between 200 and 2,000 members of Brazil's landless movement (MST), with a membership of one million, were reported to have raided a farm belonging to Monsanto. It was the third such action against Monsanto assets that year. The organisation said the property was being used to propagate seeds in readiness for the possible commercialisation of GM in Brazil. MST leader Luiz Afonso Arantes said: 'It's an illegal centre. They might be producing seeds just for research, but they are also planting with the intention of reproducing.'[33]

## PEOPLE UNITE AGAINST GM CROPS

Anti-GM activists in New Zealand have vowed to decontaminate any GM crops grown after the 2003 expiry of the moratorium on GM plantings. Lenka Rochford of the People's Moratorium Enforcement Agency (PMEA) articulates the frustration: 'We've done petitions. We've done submissions. We've written letters to the editor. We've even got naked on Parliament Grounds. We've done it all, and nobody's listening. [Direct action] is all we've got left.' Activist Penny Bright of Auckland, said: 'When injustice becomes law, resistance becomes duty. If you plant it, we'll pull it.'[34]

Similarly in the UK, the Green Gloves Pledge invited a nation of gardeners to put their gardening gloves on and take GM crops out of the ground, if they were commercialised. Some 3,300 people had signed up by Spring 2004. Happily, though, Bayer withdrew the only GM crop that had been commercialised (and that was in the GM commercialisation pipeline for the next few years) and the pledge was closed.

Japanese consumers have been actively opposing GMOs since 1996, achieving some memorable victories, particularly the December 2002 campaign to halt Monsanto's GM rice in Aichi prefecture, which resulted in the private sector abandoning GM rice R&D in Japan. This was followed by a comparable success in Iwate prefecture in November 2003, with the director of the Agriculture Department stating that no more outdoor GM research would take place in Iwate, after he received more than 407,000 signatures against it. GM research in general is getting harder all the time in Japan.[35]

All Hawaii's coffee growers have united to stop GM coffee being introduced into the state, with a joint letter and resolution opposing the growing and testing of GM coffee sent to the Hawaii Department of Agriculture.[36]

The people of Mendocino County, California, voted in a March 2004 ballot to become the first region in the US to outlaw the growing of GM crops. They won despite their opponents outspending them by a factor of six to one and raising over $600,000, most of it from CropLife America, a trade and lobbying group representing giant GM seed producers, including Monsanto, DuPont, and Dow.[37] Also in California in 2004, Trinity and Marin Counties voted to become GM-free,[38] as did Arcata City; three other counties voted, but were not successful.[39] In New England, by April 2005, nearly a hundred towns had passed resolutions to limit GM crops.[40]

### FINANCIAL PROBLEMS FOR THE BIOTECH INDUSTRY

The Institute of Science in Society (ISIS) reports that

Biotech shares peaked in 2000, but have been falling sharply since, ... performing well below the industrial average on both sides of the Atlantic. Thousands have lost their jobs in mass layoffs from the genomics ... sector. Many companies are reporting double-digit losses.[41]

Profitability remains years away.

In the US, the biotech industry clocked up a truly spectacular collective net loss of $11.6 billion in 2002 – an impressive 71 per cent increase over the $6.8 billion net loss of 2001.[42] Up to 2005, public biotech companies had clocked up a $46 billion cumulative net loss.[43] Only about 20 of the 318 public US biotech companies have had any sort of sustained profitability in the three years to 2003.[44]

Similarly in Europe, the biotech sector registered a decrease in both revenues and employee numbers in 2002. Investments for research and development fell by 11 per cent.[45] An article in *Nature Biotechnology*, May 2003, sounded the death-knell for GMOs in Europe. Here are some excerpts:

- 'Field trials of ... GM ... crops in the ... EU ... have plummeted by 87% since 1998, according to a European Commission (EC) Investigation.
- 'Two-thirds of large agroscience companies have cancelled at least one GMO research project during the same period...

- 'Most Europeans consider GM foods "of little value and dangerous for society," the Eurobarometer survey found...
- 'We are beginning to see early-stage research in Europe moving overseas and I expect that to continue.'[46]

As ISIS reports: 'Monsanto has been teetering on the brink of collapse since the beginning of 2002 as one company after another spun off their agricultural biotechnology. It has suffered a series of setbacks: drastic reductions in profits, problems in selling GM seeds in the US and Argentina,'[47] and the departure of the CEO at the end of 2002.[48] In October 2003, it started to pull out of Europe, cut up to 9 per cent of its global workforce, reported a $188 million loss, saw a big drop in share value (as much as 6 per cent),[49] and as we have seen, in 2003–4, it pulled out of a number of major crops. Syngenta, meanwhile, deserted the top plant biotech research institute in Britain, the John Innes Centre, in 2002.[50] And in June 2004, it announced it would be pulling out of its research centre at Jealott's Hill in Berkshire.[51]

### GM WHEAT REJECTED

If GM crops are so wonderful, it's puzzling why Monsanto met with such massive resistance to its GM Roundup Ready wheat from some of the most powerful players involved. The US and Canadian National Farmers Unions, the American Corn Growers Association, the Canadian Wheat Board (CWB), organic farming groups, and more than 200 other groups lobbied for a moratorium on GM wheat.

At stake, in Canada, is $4 billion in international wheat trade. The CWB reports that two-thirds of its customers do not want to buy GM wheat, or even conventional wheat that could be contaminated with GM wheat.[52] And 87 per cent of Canadian farmers say they would not grow GM wheat if they had the option. The CWB and other farm and rural groups took out large newspaper ads urging Monsanto to withdraw its federal GM wheat application.[53] As a last resort, the CWB threatened legal action.[54]

The US Wheat Associates found in its survey of wheat buyers, millers, and users that 'there is currently an overwhelming rejection' of Roundup Ready wheat.[55] US wheat exports were worth $3.6 billion in 2002.[56] Rank Hovis said: 'I am going to ask you not to grow genetically modified wheat until we are able to sell ... the bread made ... from that wheat ... if you do grow genetically modified ... wheat, we will not be able to buy any of your wheat – neither the GM nor the conventional.'[57]

Importers from Algeria, Egypt, the European Union,[58] Japan (the largest buyer of US wheat), South Korea, China,[59] the Philippines, Indonesia, and Malaysia have unequivocally and repeatedly stated that they would not accept GM wheat.[60]

Eventually, even mighty Monsanto was forced to back down, and in May 2004, it pulled out of GM wheat globally. The *New York Times* reported that the company was 'bowing to the concerns of American farmers that the crop would endanger billions of dollars of exports'.[61] Michael Rodemeyer, of the Pew Initiative on Food and Biotechnology, said that the rejection of GM wheat by American farmers could make companies reluctant to invest in other GM crops.[62]

However, in a rather sorry but altogether predictable afterword, Monsanto continues to conduct secret GM wheat trials in Canada, breaking its pledge to abandon GM wheat testing. Canadian geneticist, Professor Joe Cummins comments: '[this] exposes the absolute subservience of Canadian regulators to Monsanto'.[63]

## A WORD OF CAUTION

Despite these recent victories – particularly the dropping of new crop candidates – there is little room for complacency. Some battles may have been won, but the war is far from over. The biotech corporations have invested billions so far, and we can be sure they will not give in until their last satisfying breath.

Sue Mayer, of NGO Gene Watch, discussed in a May 2004 *Guardian* article whether we are seeing the beginning of the end of GM:

the use of GM feed for animals is likely to continue or increase as a largely "invisible" use... It is also likely that there will be attempts to use GM crops for non-food uses – including as sources of biofuels, industrial chemicals or for amenity grasses.

She goes on to point out that the developing world is

likely to form the immediate focus of the biotechnology industry's market aspirations. Pushing GM cotton into India as a bridgehead into the vast cotton markets of Asia was one step. South Africa is being used as the way into the African continent.[64]

Africa is an important target for the biotech and agrochemical corporations, since the market is so huge and because most African farmers still save their own seed and farm organically. There is also South-East Asia, where companies are bent on reaching GM corn commercialisation as swiftly as possible.[65] GM food aid is another crucial front for the GM corporations to force their products on the developing countries.[66]

Behind the old Iron Curtain, both Bulgaria and Romania already grow GM crops commercially, making them a potential cause of GM contamination for the whole region.[67] And there is virtually no legislation in all the countries of the Newly Independent States of the former Soviet Union (NIS). Monsanto and Pioneer Hi-Bred/DuPont have already exploited this to encourage extensive deployment of GM crops.[68] There is also the possibility of unlabelled GM soy and maize entering the EU through the accession states, where the borders with non-EU countries are susceptible to illicit imports.[69]

# 9

# A More Constructive Way Forward

## AT A MACRO LEVEL

I don't think any of us would disagree that, if an alternative exists to a GE solution, it's to be preferred. – Mr Hodson QC acting on behalf of the Life Sciences Network at the New Zealand Royal Commission on Genetic Modification.[1]

Plenty of better ways do exist for improving food crops without wasting millions upon millions on a technology that will only benefit the biotech corporations and which keeps thousands of scientists tied up in dead-end research. In fact, alternative forms of breeding usually yield better than GM and at a fraction of the cost. Furthermore, these methods do not involve the tyrannical patents attached to GM crops, which will turn farmers worldwide into serfs working for a handful of all-powerful agbiotech corporations. Then there is organic and sustainable agriculture, which has already delivered huge benefits to communities in the developing world. This is the real answer to hunger in the poor countries. However, the biotech lobby is not interested in this, because the only beneficiaries are the farmers, local economies, consumers, and the environment. After all, where's the sense in that? Whereas chemical-intensive and GM agriculture benefits the agbiotech corporations and the governments they sponsor, not to mention the onward march of globalisation that makes the developing countries ever more dependent on the rich nations.

### Alternative forms of breeding

#### Non-GM biotechnology

Non-GM biotechnology has enormous potential, particularly MAS, which is sometimes more loosely known as 'genomics'. Genomics is the analysis of genes and their functions in an organism, and it provides valuable information to enhance conventional breeding.[2] Over the past decade, scientists have discovered that crops are full of dormant characteristics. Rather than resorting to GM, it's often possible to simply turn on a plant's innate ability. These methods are largely uncontroversial and unpatentable.

A DNA marker sticks to a particular region of a chromosome, allowing researchers to zero in on the genes responsible for a given trait. Using this technique, much of the early stage breeding can be done in the lab, saving the time and money required to grow several generations in the field. And, unlike conventional breeding, it doesn't result in hybrids that won't come true to seed. It brings the potential to get back useful traits that have been bred out of plants over the years, like increasing the nutritional variety of vegetables, increasing disease resistance, making plants taste better, and so on.

According to a report on the subject:

In China, researcher Deng Qiyun used molecular markers while crossbreeding a wild relative of rice with his country's best hybrid to achieve a 30% jump in yield – an increase well beyond anything gained during the Green Revolution. In Bangalore, H.E. Shashidhar has cataloged the genes of the dryland varieties and used DNA markers to guide the breeding toward a high-yield super-rice, that will provide much better yields on unirrigated land. In West Africa ...breeders have created Nerica, a bountiful rice that combines the best traits of Asian and African parents. Nerica spreads profusely in early stages to smother weeds. It's disease-resistant, drought-tolerant, and contains up to 31% more protein than either parent.[3]

The head of global plant breeding at Monsanto admits: 'It's a numbers game and ultimately [non-GM] biotech [MAS] offers the greatest potential ...'[4] GM is simply not adept at dealing with complex genetic interactions like those required for drought tolerance. This is no doubt why the UK's Arable Research Institute Association and Syngenta are developing drought tolerance in sugar beet using MAS, not genetic engineering. MAS is expected to increase wheat and other non-cereal yields at more than double the forecast rate.[5] In the words of Professor Bob Goodman, former head of R&D at Calgene: 'From a scientific perspective, the public argument about genetically-modified organisms, I think, will soon be a thing of the past. The science has moved on and we're now in the genomics era.'[6]

### Traditional breeding

Farmer-developed traditional rice varieties can supply all the special traits claimed for the absurdly hyped GM varieties, except they have actually been tried and tested over hundreds of years. Scores of indigenous rice varieties are able to withstand the severest of climatic conditions, including tolerance to flooding, drought, and salinity. Here are a few examples:

- West Bengal on its own grows 78 varieties of drought-tolerant rice, Uttararnchal has 54 drought-resistant varieties, and Kerala around 40.
- Indian farmers have developed a rice that can be submerged in water for 12–15 days, while two to three days will kill conventional rice.
- In West Bengal, three varieties are grown in tidal mangrove waters, which are able to tolerate up to 14 per cent salinity. Orissa, Kerala, and Karnataka also grow many saline-resistant cultivars of rice.[7]

### Organic and sustainable agriculture

Organic agriculture avoids all agrochemicals (pesticides, insecticides, herbicides, molluscicides, etc.) and artificial fertilisers. It does not allow any GM ingredients. Livestock are allowed to roam freely and are reared without routine use of antibiotics, growth promoters, or other drugs. Clearly, then, organic agriculture is supremely more environmentally sustainable than GMOs.

In developed countries, yield reductions are minimal or zero, with yields progressing over time. In the developing countries, low-input, agro-ecological, or organic systems are commonly producing three- or four-fold yield increases. And there are manifold other advantages, too, such as improvements to soil fertility, health, and the environment, increased carbon sequestration in the soil, reduced food miles, farmer self-sufficiency, and economic and social benefits for local communities.[8]

Professor Jules Pretty has shown that nearly 9 million farmers in Africa and Asia have taken up sustainable farming on approximately 29 million hectares (an area bigger than the UK). Over the last 20 years, in 45 such projects in 17 African countries, about 730,000 households have significantly increased food output, with cereal yields improving by 50–100 per cent in the vast majority of schemes. Ethiopia, an area once reliant on emergency food aid, is able to feed itself and produce surplus crops for sale at local markets.[9] And a trial of age-old Amazonian organic methods boosted outputs on impoverished rainforest soil by a massive 880 per cent over plots using industrial agriculture.[10]

Even an editorial in the *New Scientist* concluded:

For some, talk of 'sustainable agriculture' sounds like a luxury the poor can ill afford. But in truth it is good science, addressing real needs and delivering real results. ... It is time for the major agricultural research centres and their funding agencies to join the revolution.[11]

## What should be done

If you look at the simple principle of genetic modification it spells ecological disaster. There are no ways of quantifying the risks... The solution is simply to ban the use of genetic modification in food. – Dr Harash Narang, microbiologist and senior research associate at the University of Leeds, who originally pointed to the possible link between BSE and CJD in humans[12]

In other words, we should ban the following: all GM crops; all existing GM foods, derivatives, additives, and enzymes; GM food supplements and vitamins; all GM animal feeds; all imports of GMOs; the release of all GMOs (on field trials and commercial crops) and GM farm animals; and the patenting of genetic resources for food and farm crops. We should also initiate a global review of the needs of world agriculture and food production.

## WHAT YOU CAN DO

### A call to action

- Above all, if you want to avoid GM, and you want to be part of the solution (making farming nature- and human-friendly again), eat organic foods. Organic agriculture does not allow any GM ingredients. In the UK, look for the Soil Association symbol, or a label with a UKROFS registration number. Outside the UK, buy organic food bearing a registered organic symbol.
- Ideally, avoid shopping in supermarkets, as they tend to kill local agriculture and small businesses, as well as reducing the number of local jobs and a host of other ills.[13] Try to shop in local shops, farm shops, or use an organic box scheme.
- Join an anti-GM NGO. Become involved in their GM campaigns. (See Resources for details.)
- Subscribe to an anti-GM newsletter. (See Resources for details.)
- In elections, vote for a party with a strong environmental commitment.
- Tell your friends and family what you know about GMOs. Buy them a copy of this book!

## How to avoid GMOs

### Supermarket own-brand produce

Since 1999, all the UK supermarkets have pledged to remove GM food or ingredients from all their own-brand products. On the one hand, this was great step forward for democracy; on the other, it is a bit of a farce because in practice many own-brand processed foods are likely to contain products from GM-fed animals, as indeed are their own-brand animal products. This is despite the fact that they promised some years ago to stop this practice. The various UK supermarkets have different records; in order of the best performance, they are:

- M&S: all meat, eggs, fish, and fresh milk are guaranteed non-GM fed. Other dairy products are not.
- Co-op: chicken, pork, 50 per cent of lamb, eggs, and fish (salmon and trout only) are guaranteed non-GM fed.[14] Dairy products are from non-GM fed cattle, unless labelled otherwise.[15]
- Sainsburys: chicken, eggs, and 15 per cent of pork are guaranteed non-GM fed. The company also bowed to consumer pressure, in summer 2004, and now stocks a trial range of GM-free milk in 190 stores.
- Safeway: [now owned by Morrisons].
- Waitrose: chicken, 5 per cent of pork, NZ (only) lamb, Blacktail (only) eggs, and fish are guaranteed non-GM fed.
- Asda: chicken, eggs, fish are guaranteed non-GM fed.
- Somerfield: chicken, eggs, fish are guaranteed non-GM fed.
- Tesco: chicken, eggs, fish (no for Tilapia) are guaranteed non-GM fed.[16]
- Iceland: chicken, eggs, and farmed fish (salmon and trout) are guaranteed non-GM fed.[17]
- Morrisons: nothing is guaranteed non-GM fed.[18]
- For a more up-to-date rundown on how supermarkets are doing, see: http://www.greenpeace.org.uk/Products/GM/supermarkets.cfm.

In the US, there have been no moves either to introduce labelling or to segregate GM and non-GM crops. Avoiding GMOs is therefore a real minefield. However, as we have seen, supermarket chains Wild Oats (110 stores in 22 states and British Columbia), Trader Joe's (stores in 13 states), and Whole Foods (103 stores in 22 states and Washington DC) have undertaken to eliminate GM ingredients from their own-brand products.[19] In the rest of American supermarket chains there is no such commitment, so to avoid GMOs you need to check out or print a GM-free shoppers guide from: http://www. truefoodnow.org/shoppersguide/guide_printable.html. This lists the GM status of hundreds of items found on supermarket shelves.

To find out the policies of other supermarkets around the world, go to: http://users.skynet.be/sky39402/retail.htm, which lists the website addresses of 273 grocery retailers.

### Products from GM-fed animals

In the EU, avoid products from GM-fed animals, as most non-organic meat, fish, dairy products, and eggs will be from GM-fed animals and will not be labelled. Don't forget, this includes these ingredients in processed foods, many of which contain dairy products and eggs.

### GM derivatives

In the EU, despite the April 2004 labelling regulations, there is still a 0.9 per cent threshold below which GM derivatives do not have to be labelled. Although this amounts to a very small quantity of the product, if you want to avoid GMOs completely, or the possibility of negligence or contamination, it would pay to avoid processed foods.

Soy and maize/corn derivatives are found in 80 per cent of all processed food.[20] To a lesser extent, so are rapeseed or cottonseed derivatives. These four crops are the most likely sources of GMOs. Examples of derivatives are lecithin, soy or maize oil, soy proteins, corn syrup, corn starch, vegetable oil or fat, etc.

### Other pointers

- Many brands of dairy products, cereals, jam, fruit juice, cooking oil, sweeteners, slimming foods, beverages, wine and beer, etc. are now produced with GM-enzymes. If in doubt, contact the manufacturer for assurance.
- Many brands of cheese, including vegetarian, are produced with a GM-enzyme called chymosin.
- Many additives also contain GMOs, for example:
  Riboflavin (Vitamin B2 or E101/E101A) can be produced from GMOs. Riboflavin is used in baby foods, breakfast cereals, soft drinks, slimming foods, etc.
  Caramel (E150) and xanthan gum (E415) may be derived from GM maize.

Other E numbers that may be GM-derived are: E153, E160d, E161c, E306–9, E471, E472a, E473, E475, E476b E477, E479a and b, E570, E572–3, E620–2, E624–5, E633.[21]

- Food supplements, vitamins, and medicines can contain GMOs. Contact the manufacturer to inquire about specific products.
- The UK Vegetarian Society has banned the use of its approval symbol on products containing GMOs, except GM vegetarian cheeses.
- In the UK, download and print a copy of *The Greenpeace Shopper's Guide to GM* – probably the best guide on which products contain GMOs. Products are listed according to food type, and the guide is constantly updated. Go to: http://www.greenpeace.org.uk/products/GM/goshopping.cpm

# 10
# A Last Word

This book has, by now, given ample evidence of a world with completely upside-down priorities. Big Business, as we have seen, has its claws in everything, from politicians to legislators, from scientists to journalists, from government advisory committees to international agencies, from science funding bodies to scientific journals. Bad science is routinely presented as evidence of the efficacy and safety of GM, whereas good GM-critical science is constantly ridiculed and quashed. Public interest is constantly overridden by commercial interests. Politicians seem only too eager to let all this happen. And much of the media appear to present only a business-friendly version of events.

In short, Big Business is truly out of control – or probably, more to the point, it is truly in control.

But perhaps there is a silver lining to the GM cloud, after all – which is that for all its crass injustices, its cruel plots, its conniving tricks, and its corrupt science, the biotech lobby has given us one thing – the clear crystallisation of the growing corporate control of the world, in all its callous and calculated glory.

And this is a great gift. In being so brazen and extreme, the establishment has shown us quite clearly what it is up to and how it is doing it. That is, if we choose to look. This is helping to galvanise resistance all over the world. Now, it is up to all of us to refuse to play their game, to oppose wherever we can.

Consumer boycotts of products and companies are a powerful weapon. After all, their life-blood is our money. Without it, they would collapse in no time. Writing letters to object to products or activities sends a strong message, if enough of us do it. Above all, we need to campaign for better government, to hold the corporations to account. As George Monbiot observes:

A political system is only as good as the capacity of its critics to attack it. They are the people who enforce the checks and balances which prevent any faction ... from wielding excessive power. ... We must, in other words, cause trouble. We must put the demo back into democracy. ... Legitimate protest takes many forms, including parliamentary opposition, lobbying by constituents and pressure groups, campaigning journalism and adamantly non-violent direct action. ... Troublemaking ... forces our representatives to listen to those they have failed to represent.

In fact, as we have seen in this book, agitation has already had some impressive effects.[1] Another landmark victory, worthy of note, happened in November 1999, in Seattle, when the WTO hit a wall of popular protest and demands from developing countries. This decline continued in September 2003 at the WTO Ministerial in Cancun, which collapsed over agricultural subsidies,[2] significantly undermining 'the legitimacy of the WTO', according to Alexandra Wandel of Friends of the Earth Europe.[3] The WTO is one of the most powerful of all international bodies, and until a few years ago it looked unstoppable, which proves, once again, that popular protest does work.

As a last thought, I shall leave you with the powerful and timely words of Edmund Burke (political philosopher, 1729–97): 'All that is necessary for evil to succeed is for good men to do nothing.'[4]

# Notes to the Text

(All URLs last accessed December 2004
or more recently.)

## CHAPTER 1 INTRODUCTION

1. http://www.mindfully.org/GE/GE4/Heartbreak-In-The-Heartland 21jul02.htm Percy Schmeiser, and others, speaking at 'Genetically engineered seeds of controversy: Biotech bullies threaten farmer and consumer rights', University of Texas at Austin, 10 October 2001.
2. BBC Radio Four, *Today Programme*, 28 May 2003.
3. http://www.gmwatch.org/archive2.asp?ArcId=223 *The Times*, 15 January 2003.
4. John Pilger, *Hidden Agendas* (Vintage, 1998), p. 70.
5. John Madeley, *Food for All* (Zed Books, 2002), p. 3.
6. http://www.cdc.gov/ncidod/sars/faq.htm US Centers for Disease Control and Prevention website, 'Frequently asked questions about SARS'.
7. 'HIV/AIDS in Africa', Concern literature, 2003.
8. Arthur Miller, cited in Pilger, *Hidden Agendas*, p. 44.
9. George Monbiot, *Captive State: The Corporate Takeover of Britain* (Pan, 2000), p. 17.

## CHAPTER 2 AN OVERVIEW

1. http://www.gmwatch.org/archive2.asp?ArcId=568 Letters, *Boston Globe and Mail*, 26 October 2002.
2. http://www.gmwatch.org/p1temp.asp?pid=3&page=1
3. http://www.btinternet.com/~clairejr/Pusztai/puszta_4.html *GM-Free* website.
4. Barry Commoner, 'Unravelling the DNA Myth: The spurious foundation of genetic engineering', *Harper's*, February 2002, cited in Jeffrey M. Smith, *Seeds of Deception* (Yes! Books, 2003), p. 57.
5. Commoner, 'Unravelling the DNA Myth', cited in Smith, *Seeds of Deception*, p. 53.
6. Communication with author.
7. Soil Association website, 2003.
8. Lester R. Brown et al., *State of the World 1988 – a Worldwatch Institute Report on Progress Toward a Sustainable Society* (Norton, 1988), cited in Andrew Rees, *The Pocket Green Book: The Environmental Crisis in a Nutshell* (Zed Books, 1991).
9. Helena Paul and Ricarda Steinbrecher, *Hungry Corporations: Transnational Biotech Companies Colonise the Food Chain* (Zed Books, 2003), pp. 4–5.
10. *Ibid.*, pp. 6–12.
11. *Ibid.*, pp. 1–2.

12. 'Campaigner's Guide, Edinburgh 2 July 2005, Make Poverty History', booklet.
13. http://www.oxfam.org.uk/what_we_do/issues/debt_aid/g7_deal.htm Oxfam, CAFOD, ActionAid report, 'Do The Deal: The G7 Must Act Now To Cancel Poor Country Debts', February 2005.
14. 'Campaigner's Guide, Edinburgh 2 July 2005, Make Poverty History', booklet.
15. Paul and Steinbrecher, *Hungry Corporations*, pp. 15–18.
16. *Ibid.*, pp. 23–5.
17. Andy Rowell, *Don't Worry, It's Safe to Eat: The True Story of GM Food, BSE, and Foot and Mouth* (Earthscan, 2003), p. 3.
18. *Ibid.*, p. 3.
19. Graham Harvey, *The Killing of the Countryside* (Vintage, 1997), pp. 124–5.
20. Monbiot, *Captive State*, p. 179.
21. Paul and Steinbrecher, *Hungry Corporations*, p. 79.
22. *Ibid.*, p. 18.
23. http://www.gmwatch.org/archive2.asp?arcid=888 ActionAid press release on its new report, 'GM Crops – Going Against the Grain', 28 May 2003.
24. Paul and Steinbrecher, *Hungry Corporations*, pp. 86–7.
25. *Ibid.*, p. 83.
26. www.buav.org British Union for the Abolition of Vivisection website.
27. Luke Anderson, *Genetic Engineering, Food, and Our Environment: A Brief Guide* (Green Books, 1999), p. 40.
28. India Committee of the Netherlands (ICN) press release, 4 October 2004.
29. http://www.gmwatch.org/archive2.asp?ArcId=83 *Toronto Star*, 25 February 2003.
30. *Independent*, 25 September 2003.
31. Genetic Engineering Network leaflet.
32. http://www.btinternet.com/~nlpwessex/Documents/gmoquote.htm Malcolm Walker, Chairman and Chief Executive of Iceland Foods, 26 December 1996.
33. Paul and Steinbrecher, *Hungry Corporations*, p. 173.
34. http://www.gmwatch.org/profile1.asp?PrId=43&page=F
35. ABC News poll, reported ABC News, 15 July 2004.
36. http://www.gmwatch.org/archive2.asp?arcid=1156 A survey conducted by Zhongshan University, December 2002.
37. Canada NewsWire, 8 September 2003.
38. Paul and Steinbrecher, *Hungry Corporations*, p. 173.
39. *Independent*, 25 September 2003.
40. http://www.gmwatch.org/archive2.asp?arcid=1072 'A study from the Pew Research Center', ABCNews.com, 15 July 2003.
41. http://www.gmwatch.org/archive2.asp?arcid=1156 A survey conducted by Zhongshan University, December 2002.
42. http://www.gmwatch.org/archive2.asp?arcid=1555 *Herald* DigiPoll, late August 2003.

43. http://www.gmwatch.org/archive2.asp?arcid=2637 An internet survey carried out by the Australian Consumers Association (ACA), reported on 12 February 2004.

44. http://www.gmwatch.org/archive2.asp?arcid=4835 'Canadians suspicious of biotech foods despite lack of evidence suggesting harm', Canadian Press, 25 January 2005, referring to a Pollara study from March 2004.

45. Mark Townsend, 'Supermarkets tell Blair: We won't stock GM', *Observer*, 8 June 2003.

46. Smith, *Seeds of Deception*, p. 153.

47. Correspondence with Jochen Koester, TraceConsult, 8 March 2005.

48. http://www.gmwatch.org/archive2.asp?ArcId=556 Institute of Science in Society (ISIS) condemns Prime Minister's Scoping Note, 25 October 2002.

49. http://www.organicconsumers.org/supermarket/protests0608.cfm

50. http://www.gmwatch.org/archive2.asp?arcid=4891 Kirsty Needham, 'Poultry giants quail at gene food protests', *Sydney Morning Herald*, 11 February 2005.

51. *Waikato Times*, 2nd week January 2005.

52. http://www.gmwatch.org/archive2.asp?ArcId=557 'Biotech debacle in four parts', Mae-Wan Ho and Lim Li Ching, special briefing paper for Prime Minister's Strategy Unit, 24 October 2002.

53. Anderson, *Genetic Engineering, Food, and Our Environment*, p. 120.

54. http://ngin.tripod.com/farming.htm

55. Soil Association report, 'Seeds of Doubt, North American Farmers' Experiences of GM Crops', 2002, p. 43.

56. *Ibid.*, p. 45.

57. http://ngin.tripod.com/farming.htm

58. just-food.com, 20 August 2003.

59. http://www.gmwatch.org/archive2.asp?ArcId=502 Reuters, Food Navigator (US), 10 October 2002.

60. http://ngin.tripod.com/farming.htm

61. http://www.gmwatch.org/archive2.asp?arcid=1136 Greenpeace China, 18 July 2003.

62. http://www.gmwatch.org/archive2.asp?arcid=4630 'Russian baker confirms GM-free status', CEE-foodindustry.com, 16 November 2004.

63. Anderson, *Genetic Engineering, Food, and Our Environment*, p. 119.

64. http://www.gmwatch.org/archive2.asp?ArcId=556 Institute of Science in Society (ISIS) condemns Prime Minister's Scoping Note, 25 October 2002.

65. Mark Townsend, 'Supermarkets tell Blair: We won't stock GM', *Observer*, 8 June 2003.

66. 'In the first public survey of farmers' attitudes across Australia toward GM crops (Aug 2003), Biotechnology Australia found ...', News.ninemsn, August 2003.

67. http://www.gmwatch.org/archive2.asp?arcid=4925 Mika Omura, 'Seeds of dispute: Crop crusaders', *The Asahi Shimbun*, 25 February 2005.

68. http://www.gmwatch.org/archive2.asp?arcid=1362 *StarPhoenix* (Canada), 9 August 2003.

69. www.btinternet.com/~nlpwessex/Documents/usdagmeconomics.htm

70. Smith, *Seeds of Deception*, p. 154.

71. http://www.gmwatch.org/archive2.asp?arcid=4574 'News from the American Corn Growers Foundation', press release, 27 October 2004.

72. http://www.gmwatch.org/archive2.asp?ArcId=314 Damian Wroclavsky, Reuters, 12 November 2002.

73. http://www.gmwatch.org/archive2.asp?arcid=1660 Elizabeth Johnson, *Chemical News & Intelligence*, 17 October 2003.

74. *Ecologist*, vol. 32, no. 8, October 2002.

75. http://www.gmwatch.org/archive2.asp?arcid=4180 1) The Biotech Industry organisation's website shows that the US Government organised a meeting between the Vatican and BIO 'to discuss ... the potential of biotechnology to ... ease hunger in developing countries'. http://www.bio.org/speeches/pubs/milestone04/foodandag.asp 2) The US Government got Archbishop Martino to attend its big GM promotional in Sacramento. Jennifer Garza, 'Pope to receive report on genetically altered foods', *Sacramento Bee*, 23 June 2003.

76. http://www.gmwatch.org/archive2.asp?arcid=1289 John Hooper and John Vidal, *Guardian*, 14 August 2003.

77. http://www.gmwatch.org/archive2.asp?arcid=1245 Harvey Shepherd, 'Seeds of discontent', *Montreal Gazette*, 30 November 2002.

78. http://www.gmwatch.org/archive2.asp?arcid=4359 Comments from Brother David Andrews, CSC, Executive Director, The National Catholic Rural Life Conference, Des Moines, September 2004.

79. http://www.gmwatch.org/archive2.asp?ArcId=1714 A joint position paper from the Commissioners for Environmental Questions in the Protestant Regional Churches in Germany, Commissioners for Environmental Questions in the Roman Catholic Dioceses of Germany, Protestant Services for Rural Mission, Catholic Rural Peoples' Movement, Güstrow, 7 October 2003.

80. http://www.gmwatch.org/archive2.asp?arcid=2839 Commons Hansard Debates text, 8 March 2004.

81. http://www.gmwatch.org/archive2.asp?arcid=2869 *Scotsman*, 10 March 2004.

82. http://www.gmwatch.org/archive2.asp?arcid=5227 'Developing countries face challenges by genetically modified organisms: UN report', xinhuanet.com, 11 May 2005. (Figures from UN report.)

83. http://www.gmwatch.org/archive2.asp?arcid=4845 'GM crops industry has stalled', GeneEthics Network press release, 25 January 2005.

84. http://www.gmwatch.org/archive2.asp?arcid=5227 'Developing countries face challenges by genetically modified organisms: UN report', xinhuanet.com, 11 May 2005. (Figures from UN report.)

85. http://www.gmwatch.org/archive2.asp?arcid=4845 'GM crops industry has stalled', GeneEthics Network press release, 25 January 2005.

86. http://www.gmwatch.org/archive2.asp?arcid=5227 'Developing countries face challenges by genetically modified organisms: UN report', xinhuanet.com, 11 May 2005. (Figures from UN report.)

87. Soil Association, 'Seeds of Doubt', p. 9.

88. Monbiot, *Captive State*, p. 253.

89. Environmental New Service, 16 December 2002.
90. http://www.gmwatch.org/archive2.asp?arcid=494 Elizabeth Weise, 'Letter warns Oregon about ballot measure', *USA Today*, 9 October 2002.
91. http://www.gmwatch.org/archive2.asp?ArcId=943 Claire Hope Cummings, MA, JD, CropChoice guest commentary, 11 June 2003.
92. http://www.btinternet.com/~nlpwessex/Documents/gmoquote.htm
93. *New York Times*, 25 January 2001.
94. 'Gene genie', *Sunday Herald* online, 20 July 2003.
95. Farmers Weekly Interactive, 7 July 2003.
96. Rowell, *Don't Worry*, p. 124.
97. http://www.gmwatch.org/archive2.asp?arcid=1131 Review of published studies of the health effects of GM food/feed, from Gundula Azeez, Policy Manager, Soil Association, July 2003.
98. I.F. Pryme and R. Lembcke, *Nutrition and Health*, vol. 17, 2003, pp. 1–8.
99. http://www.gmwatch.org/archive2.asp?arcid=1131 Review of published studies of the health effects of GM food/feed, from Gundula Azeez, Policy Manager, Soil Association, July 2003.
100. *Ibid.*, referring to Netherwood et al., 'Assessing the Survival of Transgenic Plant DNA in the Human Gastrointestinal Tract', *Nature Biotechnology*, vol. 22, 2004, pp. 204–9.
101. http://www.actionbioscience.org/biotech/pusztai.html Arpad Pusztai, 'Genetically modified foods: are they a risk to human/animal health?', ActionBioscience.org original article, June 2001.
102. http://www.gmwatch.org/archive2.asp?arcid=2712 New research on survival of CaMV promoter in rat tissues, Dr Terje Traavik, February 2004.
103. http://www.gmwatch.org/archive2.asp?arcid=915 Referring to S.W.B. Ewen and A. Pusztai, 'Effects of Diets Containing Genetically Modified Potatoes Expressing *Galanthus nivalis* Lectin on Rat Small Intestine', *Lancet*, vol. 354, 1999b, pp. 1353–4.
104. Smith, *Seeds of Deception*, p. 12.
105. Anderson, *Genetic Engineering, Food, and Our Environment*, p. 33.
106. http://www.gmwatch.org/p1temp.asp?pid=3&page=1
107. http://www.globalcountry.org.uk/news.php?f=uk20030708p_gm.htm#PR Global Country of World Peace press release, 9 July 2003.
108. http://www.gmwatch.org/archive2.asp?arcid=505 Greenpeace report on the EU's proposed seed contamination directive, 13 October 2002.
109. http://www.gmwatch.org/archive2.asp?arcid=3798 Edie Lau, 'Seeds of doubt part five: Grocery quandary', *Sacramento Bee*, 10 June 2004.
110. Smith, *Seeds of Deception*, p. 9.
111. 'EU Environment Ministers back tougher GM food labels', Friends of the Earth press release, 10 December 2002.
112. Ben Ayliffe, GM campaigner at Greenpeace UK, communication with author, 22 September 2004.
113. http://www.gmwatch.org/archive2.asp?arcid=4832 Greenpeace press release, 25 January 2005.

114. http://www.gmwatch.org/archive2.asp?arcid=3878 Greenpeace publish the results of a secret study from the Research Center for Milk and Foodstuffs in Weihenstephan, Bavaria. Report from the German media outlet, *derStandard*, 21 June 2004, at http://derstandard.at/text/?id=1702842

115. http://www.actionbioscience.org/biotech/pusztai.html Arpad Pusztai, 'Genetically modified foods: Are they a risk to human/animal health?', ActionBioscience.org original article, June 2001.

116. http://www.gmwatch.org/archive2.asp?arcid=505 Greenpeace report on the EU's proposed seed contamination directive, 13 October 2002.

117. Soil Association, 'Seeds of Doubt', p. 45.

118. http://www.gmwatch.org/archive2.asp?arcid=866 Devinder Sharma, 'US seeks to force-feed scientific apartheid to Third World', *Business Report* (South Africa), May 2003.

119. *Genetix Update*, issue 26, Spring 2004, p. 3.

120. Paul and Steinbrecher, *Hungry Corporations*, p. 34.

## CHAPTER 3 THE PLAYERS

1. Michael Grunwald, 'Monsanto held liable for PCB dumping', *Washington Post*, 23 February 2002.

2. Fran Abrams, 'Parliament food: "Cynical" Monsanto branded public enemy number one', *Independent*, 23 March 1999, cited in Smith, *Seeds of Deception*, p. 26.

3. Paul and Steinbrecher, *Hungry Corporations*, p. 48.

4. http://www.gmwatch.org/archive2.asp?arcid=3698 'Toxic town; town of Anniston, Alabama, is contaminated due to manufacture of PCBs', CBS News Transcripts, 60 Minutes, 10 November 2002.

5. Paul and Steinbrecher, *Hungry Corporations*, p. 84.

6. http://www.corporatewatch.org.uk/profiles/biotech/monsanto/monsanto1.html

7. 1) Monsanto website. 2) http://www.gmwatch.org/archive2.asp?arcid=988 MASIPAG report on Monsanto, 'Selling Food. Health. Hope: The real story behind Monsanto corporation, Part 1', 17 June 2003.

8. http://www.corporatewatch.org.uk/profiles/biotech/monsanto/monsanto1.html & http://www.corporatewatch.org.uk/profiles/biotech/monsanto/monsanto2.html

9. http://www.corporatewatch.org.uk/profiles/biotech/monsanto/monsanto1.html

10. http://www.corporatewatch.org.uk/genetics/monsanto.htm#pharmacia

11. Syngenta website.

12. http://www.corporatewatch.org.uk/genetics/familytree/syngenta.htm

13. *Ibid*.

14. http://multinationalmonitor.org/mm2003/03december/dec03corp1.html 'The ten worst corporations of 2003', *Multinational Monitor*, vol. 24, no. 12, December 2003.

15. http://www.corporatewatch.org.uk/profiles/bayer/bayer1.html

16. Bayer website.

17. Paul and Steinbrecher, *Hungry Corporations*, p. 84.
18. http://www.corporatewatch.org.uk/genetics/bayer.htm#bayercrop
19. http://www.corporatewatch.org.uk/profiles/dupont/dupont.htm
20. http://www.corporatewatch.org.uk/genetics/monsanto.htm#pharmacia
21. 1) Advanta website. 2) http://www.corporatewatch.org.uk/profiles/biotech/advanta/advanta1.html
22. Paul and Steinbrecher, *Hungry Corporations*, p. 84.
23. http://www.dow.com/environment/ehs.html Dow Environment Health & Safety Policy.
24. 'Dow/Union Carbide merger could be a toxic combination', INFACT press release, 30 August 1999, cited in Paul and Steinbrecher, *Hungry Corporations*, p. 49.
25. 1) Jules Pretty, *The Living Land* (Earthscan, 1998), p. 133. 2) Cargill website.
26. Paul and Steinbrecher, *Hungry Corporations*, p. 58.
27. http://www.gmwatch.org/profile1.asp?PrId=66&page=I
28. http://www.gmwatch.org/profile1.asp?PrId=66&page=I Aaron deGrassi, 'Genetically Modified Crops and Sustainable Poverty Alleviation in Sub-Saharan Africa: An Assessment of Current Evidence', published by Third World Network, Africa, 24 June 2003, p. 43. 'ISAAA implies that small farmers have been using the technology on a hundred thousand hectares. Agricultural Biotechnology in Europe – an industry coalition – suggests 5,000 ha of "smallholder cotton." The survey team suggests 3,000 ha.'
29. Rowell, *Don't Worry*, p. 7.
30. http://www.capitaleye.org/inside.asp?ID=92 Vikki Kratz, 'Food fight', *Capital Eye*, 9 July 2003.
31. http://www.gmwatch.org/profile1.asp?PrId=10&page=A CropChoice. (These figures are for 2001 and are according to controller Brian Vaught.)
32. http://www.gmwatch.org/profile1.asp?PrId=96&page=N. (These figures are for 2001 and are according to spokesman Stewart Reeve.)
33. http://www.gmwatch.org/profile1.asp?PrId=10&page=A CropChoice.
34. http://www.corporatewatch.org.uk/profiles/bayer/bayer4.htm
35. Paul and Steinbrecher, *Hungry Corporations*, p. 59.
36. 'Europe Inc.', a report by Corporate Europe Observatory, 1998, cited in Paul and Steinbrecher, *Hungry Corporations*, p. 171.
37. Adam Ma'anit, 'Exposing the biotech lobby', *Link* (Friends of the Earth International magazine), 93, April/June 2000, cited in Paul and Steinbrecher, *Hungry Corporations*, p. 171.
38. http://www.corporatewatch.org.uk/profiles/biotech/monsanto/monsanto4.html
39. Paul and Steinbrecher, *Hungry Corporations*, pp. 60–1.
40. http://www.gmwatch.org/profile1.asp?PrId=170&page=A
41. http://www.gmwatch.org/profile1.asp?PrId=170&page=A Ehsan Masood, 'GM crops: A continent divided', *Nature*, 20 November 2003.

42.  http://www.gmwatch.org/archive2.asp?arcid=269 Andy Rowell, 'The alliance of science: "Independent" groups share pro-GM common ground', *Guardian*, 26 March 2003.
43.  http://www.gmwatch.org/archive2.asp?arcid=891 CropGen press release in response to ActionAid's report, 28 May 2003.
44.  http://www.gmwatch.org/p2temp2.asp?aid=44&page=1&op=3
45.  http://www.gmwatch.org/archive2.asp?arcid=269 Andy Rowell, 'The alliance of science: "Independent" groups share pro-GM common ground', *Guardian*, 26 March 2003.
46.  http://www.gmwatch.org/profile1.asp?PrId=151&page=S
47.  http://www.gmwatch.org/profile1.asp?PrId=123&page=S
48.  http://www.gmwatch.org/archive2.asp?arcid=41
49.  http://www.gmwatch.org/archive2.asp?arcid=269 Andy Rowell, 'The alliance of science: "Independent" groups share pro-GM common ground', *Guardian*, 26 March 2003.
50.  http://www.gmwatch.org/profile1.asp?PrId=142&page=I According to Claire Fox, the IoI's Director, in an interview in *The Times*: Andrew Billen, 'A prickly opinion on just about everything',17 December 2002.
51.  http://www.gmwatch.org/archive2.asp?arcid=269 Andy Rowell, 'The alliance of science: "Independent" groups share pro-GM common ground', *Guardian*, 26 March 2003.
52.  http://www.gmwatch.org/profile1.asp?PrId=136
53.  http://www.gmwatch.org/archive2.asp?arcid=269 Andy Rowell, 'The alliance of science: "Independent" groups share pro-GM common ground', *Guardian*, 26 March 2003.
54.  http://www.gmwatch.org/p2temp2.asp?aid=26&page=1&op=1 1) 'Tobacco industry efforts subverting International Agency for Research on Cancer's second-hand smoke study', Elisa K. Ong and Stanton A. Glantz, *Lancet*, vol. 355, no. 9211, 2000, p. 1253, 2) The Philip Morris Documents website, http://www.pmdocs.com/
55.  http://www.gmwatch.org/p2temp2.asp?aid=26&page=1&op=1 http://www.prwatch.org/improp/cei.html
56.  http://www.gmwatch.org/p2temp2.asp?aid=26&page=1&op=1
57.  http://www.gmwatch.org/p2temp2.asp?aid=26&page=1&op=1 1) Dennis T. Avery, 'Warning: Organic and natural foods may be hazardous to your health', Bridge News Service, 1 October 2000. 2) Karen Charman, 'Saving the planet with pestilent statistics', prwatch, 1999.
58.  http://www.gmwatch.org/profile1.asp?PrId=14&page=A
59.  http://www.gmwatch.org/p2temp2.asp?aid=47&page=1&op=1 *Ecologist*, vol. 31, no. 1.
60.  1) http://www.lobbywatch.org/profile1.asp?PrId=85 2) http://www.gmwatch.org/p2temp2.asp?aid=47&page=1&op=1 *Ecologist*, vol. 31, no. 1.
61.  http://www.gmwatch.org/profile1.asp?PrId=44
62.  http://www.gmwatch.org/archive2.asp?ArcId=129
63.  http://www.gmwatch.org/archive2.asp?ArcId=129 George Monbiot, 'Corporate phantoms', *Guardian*, 29 May 2002.

64. http://www.gmwatch.org/archive2.asp?ArcId=129 *British Medical Journal*, vol. 326, 1 March 2003, p. 463.
65. http://www.gmwatch.org/archive2.asp?ArcId=129
66. http://www.gmwatch.org/archive2.asp?ArcId=129 'Terrorists on the march – in America', *USA Today*.
67. http://www.gmwatch.org/archive2.asp?ArcId=129
68. http://www.gmwatch.org/archive2.asp?ArcId=129 George Monbiot, 'Corporate phantoms', *Guardian*, 29 May 2002.
69. http://www.gmwatch.org/archive2.asp?arcid=2901 http://www.ewg.org/briefings/acc/index.php 'Chemical industry's secret plan to attack California's anti-toxics trend. Memo calls for phoney front groups, spying on activists', Environmental Working Group press release, 20 November 2003.
70. http://www.gmwatch.org/p2temp2.asp?aid=29&page=1&op=1
71. http://www.gmwatch.org/archive2.asp?ArcId=181 'Controversial environmental author found guilty of "scientific dishonesty"', press release, 7 January 2003.
72. Rowell, *Don't Worry*, p. 154.
73. http://www.gmwatch.org/p2temp2.asp?aid=58&page=1&op=1 George Monbiot, *Guardian*, 29 May 2002.
74. http://www.gmwatch.org/profile1.asp?PrId=106
75. http://www.gmwatch.org/archive2.asp?arcid=283
76. http://www.gmwatch.org/p2temp2.asp?aid=26&page=1&op=1
77. http://www.gmwatch.org/archive2.asp?arcid=2626 Referring to Aaron deGrassi, 'Genetically Modified Crops and Sustainable Poverty Alleviation in Sub-Saharan Africa: An Assessment of Current Evidence', published by Third World Network, Africa, 24 June 2003. p. 64.
78. Pilger, *Hidden Agendas*, p. 73.
79. Paul and Steinbrecher, *Hungry Corporations*, p. 162.
80. Monbiot, *Captive State*, p. 317.
81. Madeley, *Food for All*, pp. 118–20.
82. http://www.gmwatch.org/archive2.asp?arcid=1369 Sanjay Suri, 'Corporate giants hold government strings', Inter Press Service (IPS) News Agency, 28 August 2003, referring to Friends of the Earth press release, 28 August 2003.
83. Pilger, *Hidden Agendas*, p. 216.
84. http://www.gmwatch.org/archive2.asp?ArcId=565 'Connecting the dots on GMO implications for Montana wheat growers', CropChoice news, 24 October 2002. Dan McGuire, director of the Farmer Choice-Customer First program of the American Corn Growers Association, in his speech to the 87th annual convention of the Montana Farmers Union.
85. Monbiot, *Captive State*, p. 305.
86. *Ibid.*, pp. 316–17.
87. Paul and Steinbrecher, *Hungry Corporations*, p. 103.
88. http://www.gmwatch.org/archive2.asp?ArcId=223 John le Carré, 'The United States of America has gone mad', timesonline, 15 January 2003.
89. Pilger, *Hidden Agendas*, pp. 573–4.
90. *Ibid.*, p. 608.

91. *Ibid.*, p. 3.
92. *Ibid.*, p. 3.
93. Paul and Steinbrecher, *Hungry Corporations*, p. 106.
94. http://www.gmwatch.org/profile1.asp?PrId=131
95. http://www.gmwatch.org/archive2.asp?ArcId=651 'World Bank forges ahead with transgenic crops', Pesticide Action Network Updates Service, 27 September 2002.
96. David Korten, *When Corporations Rule the World* (Earthscan, 1995), pp. 171–2, cited in Paul and Steinbrecher, *Hungry Corporations*, p. 102.
97. Korten, *When Corporations Rule*, p. 160, cited in Paul and Steinbrecher, *Hungry Corporations*, p. 103.
98. Korten, *When Corporations Rule*, p. 159, cited in Paul and Steinbrecher, *Hungry Corporations*, p. 102.
99. http://www.gmwatch.org/profile1.asp?PrId=99&page=O
100. Madeley, *Food for All*, pp. 158–9.
101. http://www.gmwatch.org/archive2.asp?arcid=946 Matt Mellen, 'Who is getting fed?' GRAIN publications, Seedling, April 2003.
102. Paul and Steinbrecher, *Hungry Corporations*, p. 208.
103. http://www.gmwatch.org/archive2.asp?ArcId=574 Afrol News, via www.africapulse.org, 30 October 2002.
104. Bertini is quoted in 'New technology "terminates" food independence', World Internet News Distribution Source (WINDS), April 1998, cited in Paul and Steinbrecher, *Hungry Corporations*, p. 210.
105. Madeley, *Food for All*, p. 157.
106. Paul and Steinbrecher, *Hungry Corporations*, p. 151.
107. http://www.gmwatch.org/archive2.asp?arcid=3627 'Agriculture: Towards 2015/30', report from the UN Food and Agriculture Organization's (FAO) Global Perspective Studies Unit, July 2002.
108. Sarah Boseley, 'WHO "infiltrated by food industry"', *Guardian*, 9 January 2003.
109. http://www.gmwatch.org/archive2.asp?ArcId=561 *Lancet*, vol. 360, no. 9342, 26 October 2002.
110. Madeley, *Food for All*, pp. 52–3.
111. Paul and Steinbrecher, *Hungry Corporations*, p. 107.
112. http://www.gmwatch.org/archive2.asp?ArcId=367 'CGIAR openly adopts corporate agenda', AgBioIndia mailing list, 5 November 2002.
113. Paul and Steinbrecher, *Hungry Corporations*, pp. 112–13.
114. http://www.gmwatch.org/archive2.asp?arcid=5232 'Industry mobilizes to modify Mexico's labeling measures', Cropchoice news, 12 February 2001.
115. http://www.gmwatch.org/archive2.asp?arcid=5228
116. http://www.gmwatch.org/archive2.asp?arcid=4852 'US pressing internationally for lax food safety testing for GM foods', Gene Campaign press release, 1 February 2005.
117. http://www.gmwatch.org/archive2.asp?arcid=4206 Sean Hao, 'USDA told to disclose "biopharm" locations', *Honolulu Advertiser*, 5 August 2004.

118. http://www.gmwatch.org/archive2.asp?arcid=3880 'US speeding food technology transfers to Africa, Veneman says', United States Department of Agriculture (Washington, DC) document, 21 June 2004.

119. Gwynne Dyer, 'Frankenstein foods', *Globe and Mail*, 20 February 1999, cited in Smith, *Seeds of Deception*, p. 150.

120. Monbiot, *Captive State*, p. 241.

121. ensnews.com, 23 June 2003.

122. http://www.gmwatch.org/archive2.asp?arcid=5205 Stephen Clapp, 'USDA launches initiative to foster specialty biotech crops', *Pesticide & Toxic Chemical News*, 21 March 2005.

123. http://www.gmwatch.org/print-archive2.asp?arcid=464 'Agribusiness takes most seats on USDA biotech panel', Reuters Securities News, 9 April 2003.

124. http://www.gmwatch.org/archive2.asp?arcid=946 Matt Mellen, 'Who is getting fed?', GRAIN publications, Seedling, April 2003.

125. http://www.gmwatch.org/archive2.asp?ArcId=392

126. http://www.gmwatch.org/profile1.asp?PrId=165&page=U

127. http://www.gmwatch.org/archive2.asp?arcid=865 Friends of the Earth International press release, 23 May 2003.

128. http://www.gmwatch.org/archive2.asp?arcid=300

129. http://www.gmwatch.org/profile1.asp?PrId=165&page=U

130. http://www.gmwatch.org/archive2.asp?ArcId=532 Food First press release, 15 October 2002.

131. *GM-Free*, vol. 1, no. 1, April 1999, p. 8.

132. Dr Sue Mayer, of GeneWatch UK, cited in Rowell, *Don't Worry*, p. 170.

133. Andy McSmith, 'Blair buried health warning on GM crops, says sacked minister', *Independent*, 22 June 2003.

134. Monbiot, *Captive State*, p. 244.

135. http://www.gmwatch.org/archive2.asp?arcid=1164 Andy Rowell, 'Debate, what debate? The GM public debate – legitimate discussion of the concerns, or political gameplaying?', *Ecologist*, GM special, July/August 2003.

136. Monbiot, *Captive State*, p. 262.

137. R.A.E. North, *The Death of British Agriculture* (Duckworth, 2001), p. 208.

138. http://www.gmwatch.org/profile1.asp?PrId=43&page=F

139. http://www.gmwatch.org/profile1.asp?PrId=73&page=K

140. http://www.gmwatch.org/archive2.asp?arcid=1164 Andy Rowell, 'The GM public debate – legitimate discussion of the concerns, or political gameplaying?', *Ecologist*, July/August 2003.

141. http://www.gmwatch.org/profile1.asp?PrId=43&page=F 'In a joint letter [March 2003] to the FSA, a number of leading UK organisations, including the National Federation of Women's Institutes and the UK's largest trade union, UNISON, condemned not just the "biased materials the FSA had created for the debate" but the FSA's complete failure to co-operate with the Government-sponsored GM Public Debate ... "There is a strong consensus amongst consumer and environment organisations," the letter said, "that the published views and statements of the FSA and

its Chair are indistinguishable from those of the pro-GM lobby and do not properly represent public health and consumer interests."'

142. http://www.gmwatch.org/archive2.asp?ArcId=362 Geoffrey Lean, 'Organic farming shunned by food watchdog', *Independent on Sunday*, 3 November 2002.

143. http://www.gmwatch.org/archive2.asp?ArcId=362 Letters, *Guardian*, 14 September 2000.

144. http://www.gmwatch.org/profile1.asp?PrId=43&page=F

145. Monbiot, *Captive State*, p. 277.

146. http://www.gmwatch.org/p2temp2.asp?aid=36&page=1&op=1

147. Monbiot, *Captive State*, p. 278.

148. Testimony of Professor Phillip James and Dr Andrew Chesson, Examination of Witnesses (Questions 207–219), 8 March 1999, cited in Smith, *Seeds of Deception*, p. 25.

149. http://www.gmwatch.org/profile1.asp?PrId=21

150. http://www.i-sis.org.uk/EngineeringLifeAndMind.php

151. Monbiot, *Captive State*, p. 275.

152. http://www.gmwatch.org/archive2.asp?arcid=1570 George Monbiot, 'Force-fed a diet of hype', *Guardian*, 7 October 2003.

153. http://www.gmwatch.org/profile1.asp?PrId=21

154. http://www.gmwatch.org/profile1.asp?PrId=175&page=B

155. http://www.gmwatch.org/archive2.asp?arcid=2291

156. Harvey, *The Killing of the Countryside*, p. 112.

157. http://www.gmwatch.org/p2temp2.asp?aid=36&page=1&op=1

158. Monbiot, *Captive State*, p. 273.

159. http://www.gmwatch.org/p2temp2.asp?aid=36&page=1&op=1

160. J. Leake, 'Royal Society found guilty of keeping out woman scientists', *The Sunday Times*, 28 July 2002, p.7, cited in Rowell, *Don't Worry*, p. 104.

161. http://www.gmwatch.org/profile1.asp?PrId=113&page=R See, for example, The Royal Society Annual Review 1998–99, p. 26.

162. http://www.gmwatch.org/profile1.asp?PrId=113&page=R

163. http://www.gmwatch.org/p2temp2.asp?aid=28&page=1&op=3

164. Rowell, *Don't Worry*, pp. 103–5.

165. *Ibid.*, pp. 120–2.

166. Dr Arpad Pusztai, cited in Rowell, *Don't Worry*, p. 103.

167. Professor Lacey, cited in Rowell, *Don't Worry*, p. 109.

168. http://www.gmwatch.org/archive2.asp?ArcId=286 Laurie Flynn and Michael Sean Gillard, 'Pro-GM scientist "threatened editor"', *Guardian*, 1 November 1999.

169. http://www.gmwatch.org/profile1.asp?PrId=113&page=R http://www.absw.org.uk/guidance_for_editors.htm Association of British Science Writers website.

170. http://www.gmwatch.org/p2temp2.asp?aid=50&page=1&op=1 1) 'Greenpeace wins damages over Professor's "unfounded" allegations,' *Education Guardian*, 8 October 2001. 2) *Ecologist*, vol. 31, no. 10, p. 11. 3) *Private Eye*, 1040, p. 4.

171. http://www.gmwatch.org/archive2.asp?ArcId=44 GM-free Cymru and partners press release, 7 February 2003.

## CHAPTER 4 EXPOSING THE WILD CLAIMS
## MADE BY THE BIOTECH LOBBY

1. http://www.gmwatch.org/archive2.asp?arcid=1525 Michael McCarthy, 'Blair the key as decision nears over commercial GM crops in Britain', *Independent*, 25 September 2003.
2. http://www.gmwatch.org/archive2.asp?arcid=921 'GM Crops: Do we need them? Are they safe?' First conference of the Independent Science Panel, King's College London, 10 May 2003.
3. http://plus.i-bio.gov.uk/ibioatlas/textd1.html UK Government's Genetically Modified Organisms (Contained Use) Regulations 2000 (CUR 2000).
4. http://www.gmwatch.org/archive2.asp?arcid=936 Quotes from members of the Independent Science Panel on GM, 11 June 2003.
5. L. Kahl, memorandum about the Federal Register Document 'Statement of policy: Foods from genetically modified plants', to James Maryanski, 8 January 1992, cited in Rowell, *Don't Worry*, p. 128.
6. M.J. Smith, 'Comments on draft Federal Register Notice on Food Biotechnology', memorandum to James Maryanski, 8 January 1992, cited in Rowell, *Don't Worry*, p. 128.
7. BBC Radio Four, *Seeds of Trouble*, 7 January 2003.
8. T. Inose, *International Journal of Food Science Technology*, vol. 30, 1995, 141.
9. Communication with author.
10. India Committee of the Netherlands (ICN) press release, 4 October 2004.
11. http://www.gmwatch.org/archive2.asp?arcid=855 NLP (Natural Law Party) Wessex bulletin.
12. http://www.gmwatch.org/archive2.asp?arcid=964 Michael Meacher, 'Are GM crops safe? Who can say? Not Blair', *Independent on Sunday*, 22 June 2003.
13. Smith, *Seeds of Deception*, p. 39.
14. http://www.gmwatch.org/archive2.asp?arcid=964 Michael Meacher, 'Are GM crops safe? Who can say? Not Blair', *Independent on Sunday*, 22 June 2003.
15. 'Scientists concerned about spread of GMOs in Russia', RBC News, 11 December 2003.
16. http://www.gmwatch.org/archive2.asp?arcid=921 'GM Crops: Do we need them? Are they safe?' First conference of the Independent Science Panel, King's College London, 10 May 2003.
17. http://www.gmwatch.org/archive2.asp?arcid=4524 'Long-term effect of GM crops serves up food for thought', *Nature*, vol. 398, p. 651.
18. http://www.btinternet.com/~nlpwessex/Documents/gmoquote.htm
19. Andrea Baillie, 'Suzuki warns of Frankenstein foods', CP Wire, 18 October 1999, cited in Smith, *Seeds of Deception*, pp. 136–7.
20. T. Inose, *International Journal of Food Science Technology*, vol. 30, 1995, p. 141.
21. 'Long-term effect of GM crops serves up food for thought', *Nature*, vol. 398, p. 651.

22. Five Year Freeze, 'Feeding or Fooling the World?', p. 10.
23. Pallab Ghosh, 'GM crops "good for developing countries"', BBC News, 10 June 2003.
24. 'Let nature's harvest continue!' response to Monsanto from delegates from 20 African countries to the FAO of the UN, cited in Five Year Freeze, 'Feeding or Fooling the World?', p. 3.
25. ActionAid report, 'GM Crops – Going Against the Grain', May 2003.
26. http://www.gmwatch.org/archive2.asp?arcid=1042 Katharine Ainger, 'Is George Bush the new Bob Geldof?', New Statesman, 30 June 2003.
27. http://www.gmwatch.org/archive2.asp?arcid=2696 Geoffrey Lean, 'GM seeds may have built-in obsolescence', Independent on Sunday, 22 February 2004.
28. http://www.gmwatch.org/archive2.asp?ArcId=488 Dennis Bueckert, 'Activists campaign against patents for seeds, see biodiversity threat', CP Wire (via Agnet), 4 October 2002.
29. 'No point to GM say academics', One News (NZ), 7 September 2003.
30. USDA report, 'The Adoption of Bioengineered Crops', June 2002. Available from: http://www.ers.usda.gov/publications/aer810/ Summary available from: www.btinternet.com/~nlpwessex/Documents/usdagmeconomics.htm
31. Soil Association report, 'Seeds of Doubt, North American Farmers' Experiences of GM Crops', 2002, p. 19. Available free from: http://www.non-gm-farmers.com/ Summary at: http://www.gmwatch.org/archive2.asp?ArcId=548
32. http://www.gmwatch.org/archive2.asp?ArcId=2000 Referring to Michael Duffy, 'Who Benefits from Biotechnology?', January 2002.
33. Soil Association, 'Seeds of Doubt', p. 20.
34. http://www.gmwatch.org/archive2.asp?ArcId=182 'GM crops are an economic disaster', Soil Association press release, 17 September 2002.
35. Soil Association, 'Seeds of Doubt', p. 19.
36. Advertisement in Top Producer, January 2002 ('Asgrow' is a trademark of Monsanto Company), cited in Soil Association, 'Seeds of Doubt', p. 19.
37. 1) http://www.gmwatch.org/archive2.asp?ArcId=2103 2) http://www.i-sis.org/GMcropsfailed.php Referring to Elmore et al., 'Glyphosate-Resistant Soybean Cultivar Yields Compared with Sister Lines', Agronomy Journal, vol. 93, 2001, pp. 408–12.
38. Ecologist, vol. 32, no. 8, October 2002.
39. Farmers Weekly (UK), 4 December 1998.
40. http://www.btinternet.com/~nlpwessex/Documents/gmlemmings.htm Bill Christison at a conference on Biodevastation, St Louis, Missouri, July 1998.
41. 1) Crop Choice News, 29 September 2001. 2) www.mslawyer.com/mssc/ctapp/20010925/0000137.html cited in Soil Association, 'Seeds of Doubt', p. 12.
42. Advertisement in Top Producer, January 2002, cited in Soil Association, 'Seeds of Doubt', p. 11.
43. 'The Roundup Ready Soyabean System: Sustainability and herbicide use', Monsanto, April 1998, cited in Soil Association, 'Seeds of Doubt', p. 12.

44. *Ibid.*, cited in Soil Association, 'Seeds of Doubt', p. 16.
45. USDA, 'The Adoption of Bioengineered Crops'.
46. Soil Association, 'Seeds of Doubt', pp. 15–18.
47. http://www.gmwatch.org/archive2.asp?arcid=1722 Referring to Dr Charles Benbrook, 'Impacts of Genetically Engineered Crops on Pesticide Use in the United States: The First Eight Years', November 2003. Available from: http://www.biotech-info.net/technicalpaper6. html
48. Soil Association, 'Seeds of Doubt', pp. 16–17.
49. *Ibid.*, p. 23.
50. *Ibid.*, pp. 16–17.
51. 'Monsanto sees opportunity in glyphosate resistant volunteer weeds', www.cropchoice.com, 3 August 2001, cited in Soil Association, 'Seeds of Doubt', p. 24.
52. Soil Association, 'Seeds of Doubt', p. 24.
53. *Ibid.*, p. 59.
54. J. Mendelson, 'The world's biggest-selling herbicide', *Ecologist*, vol. 28, no. 5, September/October 1998, cited in Soil Association, 'Seeds of Doubt', p. 59.
55. Stephen Nottingham, *Genescapes, The Ecology of Genetic Engineering* (Zed Books, 2002), cited in Soil Association, 'Seeds of Doubt', p. 59.
56. Soil Association, 'Seeds of Doubt', p. 58.
57. 'Western Canadian farmers growing more GM crops', *Leader-Post*, 20 October 2001, cited in Soil Association, 'Seeds of Doubt', p. 58.
58. http://www.gmwatch.org/archive2.asp?arcid=1054 Robert Uhlig, *Daily Telegraph*, 10 July 2003.
59. http://www.gmwatch.org/profile1.asp?PrId=150&page=W
60. 1) http://www.gmwatch.org/archive2.asp?ArcId=3906 2) http://www. fao.org/documents/show_cdr.asp?url_file=/DOCREP/003/X9602E/ x9602e02.htm UN FAO Corporate Document Repository, 'Genetically modified organisms, consumers, food safety and the environment', quotations from the English language media.
61. http://www.gmwatch.org/archive2.asp?arcid=3268 Anastasia Stephens, 'Puncturing the GM myths', *Evening Standard*, 8 April 2004, from an interview with Dr Mae-Wan Ho, Director of the Institute of Science in Society.
62. http://www.gmwatch.org/archive2.asp?arcid=1839 Ann Scholl and Facundo Arrizabalaga, 'Argentina: The catastrophe of GM soya', *Green Left Weekly*, 12 November 2003.
63. http://www.gmwatch.org/archive2.asp?arcid=3280 'Argentina's bitter harvest', *New Scientist*, 17 April 2004.
64. http://www.gmwatch.org/archive2.asp?ArcId=3906 Professor Jules Pretty, 'A strange fruit', *Red Pepper*, 10 December 2000.
65. Graham Wynne, Chief Executive of the RSPB, cited in Anderson, *Genetic Engineering, Food, and Our Environment*, p. 27.
66. Mark Townsend, 'Number 10's wildlife experts warn against GM damage', *Observer*, 19 October 2003.
67. USDA, 'The Adoption of Bioengineered Crops'.

68. http://www.gmwatch.org/archive2.asp?arcid=527 Charles Margulis, 'Playing with our food: A massive food experiment already underway; genetically engineered foods', *Earth Island Journal*, 22 December 2001.

69. Five Year Freeze, 'Feeding or Fooling the World?', p. 18.

70. Jules Pretty, *Agri-Culture* (Earthscan, 2002), p. 86.

71. http://www.gmwatch.org/archive2.asp?ArcId=475 ISIS report, 'Organic Agriculture Fights Back', 2 October 2002.

72. Five Year Freeze, 'Feeding or Fooling the World?', p. 18.

73. *Ibid.*, p. 20.

74. *Ibid.*, p. 22.

75. http://www.gmwatch.org/archive2.asp?ArcId=198 'Providing proteins to the poor – genetically engineered potatoes vs. amaranth and pulses', Research Foundation for Science Technology & Ecology, 9 January 2003.

76. Five Year Freeze, 'Feeding or Fooling the World?', pp. 20–5.

77. Opinion piece about Golden rice by Benedikt Haerlin, cited in Smith, *Seeds of Deception*, p. 210.

78. Greenpeace demands false biotech advertising be removed from TV, Letter, 9 February 2001, cited in Smith, *Seeds of Deception*, p. 210.

79. *Ibid.*, cited in Smith, *Seeds of Deception*, p. 211.

80. Five Year Freeze, 'Feeding or Fooling the World?', pp. 20–5.

81. http://www.gmwatch.org/archive2.asp?arcid=1709

82. http://www.gmwatch.org/archive2.asp?arcid=1010 Devinder Sharma, 'From Pomato to Protato, Bending it unlike Beckham'.

83. http://www.gmwatch.org/archive2.asp?arcid=1709

## CHAPTER 5 THE RISKS AND DANGERS OF GMOS

1. Soil Association, 'Seeds of Doubt', p. 25.

2. *Ibid.*, p. 58.

3. Ricarda Steinbrecher and Jonathan Latham, 'Horizontal Gene Transfer from GM Crops to Unrelated Organisms', *Econexus*, 27 January 2003.

4. http://www.gmwatch.org/archive2.asp?arcid=2906 Judith Jordan, Product Manager for Aventis asked, as a witness for the prosecution in the failed April 2000 Greenpeace court case, whether a 50-metre buffer zone between GM and conventional crops was really sufficient to prevent cross contamination, replied that pollution in those circumstances was as likely as getting pregnant from a toilet seat.

5. http://www.defra.gov.uk/news/2003/031013b.htm 'GM research reports published', DEFRA Information Bulletin, 13 October 2003.

6. http://www.gmwatch.org/archive2.asp?arcid=2623 Steve Dube, 'WAG taking time over GM decision', *Western Mail*, 10 February 2004.

7. http://www.gmwatch.org/archive2.asp?arcid=4376 Fred Pearce, 'Wind carries GM pollen record distances', *New Scientist*, NewScientist.com news service, 20 September 2004.

8. A. Dove et al., 'Research News: Promiscuous Pollination', *Nature Biotechnology*, vol. 16, September 1998, p. 805, cited in Anderson, *Genetic Engineering, Food, and Our Environment*, p. 36.

9. J. Burgelson, C.B. Purrington and G. Wichmann, 'Promiscuity in Transgenic Plants', *Nature*, vol. 395, 3 September 1998, p. 25, cited in Anderson, *Genetic Engineering, Food, and Our Environment*, p. 36.

10. Soil Association, 'Seeds of Doubt', p. 26.

11. *Ibid.*, p. 27.

12. http://www.gmwatch.org/archive2.asp?arcid=1510 Dr Peter Wills, Department of Physics, University of Auckland, 24 September 2003.

13. http://www.gmwatch.org/archive2.asp?arcid=4315 'New Research reveals widespread GMO contamination and threats to local agriculture from the world's first commercially planted genetically engineered tree', Hawaii GEAN press release, 9 September 2004.

14. http://www.gmwatch.org/archive2.asp?arcid=2290

15. http://www.gmwatch.org/archive2.asp?ArcId=372 'GM mustard will pay havoc with Indian food chain', Forum for Biotechnology & Food Security (New Delhi) press release, 6 November 2002.

16. Martin Phillipson, Professor of Law, Canadian Bar Association's annual conference, August 2001.

17. http://www.gmwatch.org/archive2.asp?ArcId=372 'GM mustard will pay havoc with Indian food chain', Forum for Biotechnology & Food Security (New Delhi) press release, 6 November 2002.

18. Soil Association, 'Seeds of Doubt', p. 22.

19. http://www.gmwatch.org/archive2.asp?arcid=325 Syngenta market research study report, circulated to farmers and landowners via its PR company, Gibbs & Soell, late 2002.

20. http://www.gmwatch.org/archive2.asp?arcid=4297 David Bennett, '"Highly suspicious" plants found in north-central Arkansas', *Delta Farm Press*, 30 August 2004.

21. http://www.gmwatch.org/archive2.asp?arcid=325 Syngenta market research study report, circulated to farmers and landowners via its PR company, Gibbs & Soell, late 2002.

22. http://www.gmwatch.org/archive2.asp?arcid=182 'USDA report exposes GM crop economics myth', NLP Wessex, 22 August 2002.

23. http://www.gmwatch.org/archive2.asp?arcid=325 Syngenta market research study report, circulated to farmers and landowners via its PR company, Gibbs & Soell, late 2002.

24. http://www.gmwatch.org/archive2.asp?ArcId=541 *Western Morning News*, 21 October 2002.

25. Geoffrey Lean 'Genetically modified strains have contaminated two-thirds of all crops in US', *Independent*, 7 March 2004.

26. http://www.gmwatch.org/archive2.asp?arcid=1525 Michael McCarthy, 'Blair the key as decision nears over commercial GM crops in Britain', *Independent*, 25 September 2003.

27. http://www.gmwatch.org/archive2.asp?arcId=61 Bob Burton, Inter Press Service (IPS) News Agency, 13 February 2003.

28. Ashok Sharma, 'GM crops: Scientific analysis needed, not hype', *Financial Express* (India), 14 February 2005.

29. http://www.soilassociation.org/web/sa/saweb.nsf/848d689047
    cb466780256a6b00298980/b506211331429c6d80256e4300536165!
    OpenDocument Soil Association website: library.
30. http://www.gmwatch.org/archive2.asp?arcid=5127 'German Consumer
    Protection Minister: "Unbelievable sloppiness!"', *Spiegel International*
    (Germany), 18 April 2005.
31. http://www.gmwatch.org/archive2.asp?arcid=1613 Indigenous &
    farming communities in Oaxaca, Puebla, Chihuahua, Veracruz
    CECCAM, CENAMI, ETC Group, CASIFOP, UNOSJO, AJAGI press
    release, 9 October 2003.
32. 'A bridge too far', Greenpeace news, from website, 18 August 2003.
33. Rowell, *Don't Worry*, p. 150.
34. 'A bridge too far', Greenpeace news, from website, 18 August 2003.
35. 'Models show gene flow from crops threatens wild plants', posted on
    the University of Wisconsin, Madison, website, 23 July 2003.
36. Soil Association, 'Seeds of Doubt', p. 39.
37. Anderson, *Genetic Engineering, Food, and Our Environment*, p. 52.
38. http://www.gmwatch.org/archive2.asp?ArcId=500 *New Scientist*, 17
    August 2002.
39. http://www.gmwatch.org/archive2.asp?arcid=2634
40. Soil Association, 'Seeds of Doubt', p. 18.
41. http://www.gmwatch.org/archive2.asp?arcid=1510 Dr Peter R. Wills,
    Department of Physics, University of Auckland, 24 September 2003.
42. http://www.gmwatch.org/archive2.asp?ArcId=355 Ranjit Devraj, 'A
    mindless conviction', Inter Press Service (IPS), October 2002.
43. http://www.gmwatch.org/archive2.asp?arcid=1661 Peter Rosset, 'UK
    study finds new risks in genetically engineered crops', IPS Columnist
    Service, October 2003.
44. http://www.gmwatch.org/archive2.asp?arcid=1353 Jeremy Bigwood,
    'GM crop weed killer linked to powerful fungus', *Counterpunch*, 23
    August 2003.
45. Barry Commoner, 'Unravelling the DNA Myth: The spurious foundation
    of genetic engineering', *Harper's*, February 2002, cited in Smith, *Seeds
    of Deception*, p. 75.
46. http://www.gmwatch.org/archive2.asp?arcid=1371 Gene Campaign
    press release (Delhi), 8 August 2003.
47. Smith, *Seeds of Deception*, pp. 47–8.
48. Soil Association, 'Seeds of Doubt', p. 36.
49. *Ibid.*, p. 37.
50. http://www.gmwatch.org/archive2.asp?arcid=1510 Dr Peter R. Wills,
    Department of Physics, University of Auckland, 24 September 2003.
51. http://www.gmwatch.org/archive2.asp?arcid=2774 John Aglionby in
    the southern Philippines, 'Filipino islanders blame GM crop for mystery
    sickness', *Guardian*, 3 March 2004.
52. Soil Association, 'Seeds of Doubt', pp. 35–6.
53. *Ibid.*, p. 36.
54. http://www.gmwatch.org/archive2.asp?arcid=1890 'Cows die
    mysteriously on farm in Hesse/Germany', Greenpeace Germany, 8
    December 2003.

55. http://www.gmwatch.org/archive2.asp?arcid=3344
56. http://www.gmwatch.org/archive2.asp?arcid=1890 'Animals avoid GM food, for good reasons', ISIS press release, 13 December 2003.
57. Smith, *Seeds of Deception*, p. 170.
58. William Ryberg, 'Growers of biotech corn say they weren't warned: StarLink tags appear to indicate it's suitable for human food products', *Des Moines Register*, 25 October 2000, cited in Smith, *Seeds of Deception*, p. 167.
59. Reuters News Service, 'StarLink tainted corn from 1999, 2000 crops', Washington, 20 March 2001, cited in Rowell, *Don't Worry*, p. 133.
60. www.gmwatch.org/archive2.asp?arcid=4016 ABC News, 28 November 2000.
61. Jonathan Bernstein, and others, 'Clinical and laboratory investigation of allergy to genetically modified foods', *Environmental Health Prospectus*, vol. 111, 2003, pp. 1114–21, cited in Smith, *Seeds of Deception*, p. 166.
62. Bill Freese, 'The StarLink affair, submission by Friends of the Earth to the FIFRA Scientific Advisory Panel considering assessment of additional scientific information concerning StarLink corn', 17–19 July 2001, cited in Smith, *Seeds of Deception*, p. 176.
63. Smith, *Seeds of Deception*, p. 176.
64. Alan Rulis, Center for Food Safety and Applied Nutrition, to Sally Van Wert, AgrEvo USA Company, 29 May 1998, cited in Smith, *Seeds of Deception*, p. 167.
65. Smith, *Seeds of Deception*, pp. 170–1.
66. FIFRA Scientific Advisory Panel (SAP), open meeting, 17 July 2001, cited in Smith, *Seeds of Deception*, pp. 170–1.
67. Smith, *Seeds of Deception*, p. 173.
68. Bill Freese, 'The StarLink affair, submission by Friends of the Earth to the FIFRA Scientific Advisory Panel considering assessment of additional scientific information concerning StarLink corn', 17–19 July 2001, cited in Smith, *Seeds of Deception*, pp. 175–6.
69. http://www.gmwatch.org/archive2.asp?ArcId=1300 'Illegal genetically engineered StarLink corn contaminates food aid; other types not approved by the EU also found', Friends of the Earth press release, 10 June 2002.
70. http://www.gmwatch.org/archive2.asp?ArcId=177 Willie Vogt, 'Japan to boost StarLink testing', *Farm Progress*, 1 June 2003.
71. 1) http://www.gmwatch.org/archive2.asp?arcid=4916 *El Diario* (La Paz, Bolivia), 12 June 2002. A Bolivian group – El Foro Boliviano para el Desarrollo ye el Medio Ambiente – criticizes the USAID for shipping food aid contaminated with StarLink to Bolivia. 2) http://www.gmwatch.org/archive2.asp?arcid=4932 Aina Hunter, 'Pharm aid: Genetically altered corn, banned for health reasons, pops up in US aid shipments to Guatemala', villagevoice.com, 28 February 2005.
72. http://www.gmwatch.org/archive2.asp?arcid=1796 Eva Cheng, 'Mexico: Campaign against GM contamination', greenleft.org.au.
73. http://www.gmwatch.org/archive2.asp?arcid=1796 Paul Jacobs, 'Traces of contaminated grain still showing up in corn supply', *Knight Ridder Newspapers*, 1 December 2003.

74. *Farmers Weekly* (UK), 8 December 2000.
75. http://www.gmwatch.org/archive2.asp?arcid=5024 'GMO crop scandal
    – did Syngenta's illegal corn come to Europe?', Friends of the Earth
    Europe press release, 23 March 2005.
76. http://www.gmwatch.org/archive2.asp?arcid=5177 Jeffrey M. Smith,
    'US government and biotech firm deceive public on GM corn mix-up',
    Spilling the Beans, Institute for Responsible Technology, April 2005.
77. http://www.gmwatch.org/archive2.asp?arcid=5053 Paul Brown, 'Joint
    US–UK cover-up alleged over GM maize', *Guardian*, 1 April 2005.
78. http://www.gmwatch.org/archive2.asp?arcid=5029 'GM maize imported
    into Europe had no US or EU approval,' press release from GM-free
    Cymru, 24 March 2005.
79. http://www.gmwatch.org/archive2.asp?arcid=5024
80. http://www.gmwatch.org/archive2.asp?arcid=5073 'GM maize scandal:
    Authorities slammed for incompetence and complacency', press release
    from GM-free Cymru, 4 April 2005.
81. http://www.gmwatch.org/archive2.asp?arcid=5346 Email from Syngenta
    to DEFRA, 5 April 2005.
82. http://www.gmwatch.org/archive2.asp?arcid=5286 Critique of
    Syngenta's documents (Syngenta Biotechnology Report SSB-104–05
    and SSB-112–05 on Bt10) by Dr Jack Heinemann, Director of New
    Zealand Institute of Gene Ecology, in an email to Claire Bleakley, 18
    May 2005.
83. http://www.gmwatch.org/archive2.asp?arcid=5053 Paul Brown, 'Joint
    US–UK cover-up alleged over GM maize', *Guardian*, 1 April 2005.
84. http://www.gmwatch.org/archive2.asp?arcid=5073
85. http://www.gmwatch.org/archive2.asp?arcid=5114 'Japan wary of
    making new purchases of US corn', Illinois Farm Bureau, 15 April
    2005.
86. http://www.gmwatch.org/archive2.asp?arcid=5145
87. http://www.gmwatch.org/archive2.asp?arcid=5100 'Illegal GE rice
    contaminates food chain in China', Greenpeace press release, 13 April
    2005.
88. http://www.gmwatch.org/archive2.asp?arcid=5118 David Barboza,
    'China's problem with "anti-pest" rice', *New York Times*, 16 April
    2005.
89. http://www.gmwatch.org/archive2.asp?arcid=5105 Paul Brown,
    'Greenpeace finds illegal strain in Chinese exports', *Guardian*, 14 April
    2005.
90. http://www.gmwatch.org/archive2.asp?arcid=5100 'Illegal GE rice
    contaminates food chain in China', Greenpeace press release, 13 April
    2005.
91. http://www.gmwatch.org/archive2.asp?arcid=5350 Xun Zi, 'GM rice
    forges ahead in China amid concerns over illegal planting', *Nature
    Biotechnology*, vol. 23, June 2005, p. 637.
92. 'Stung into action', *Guardian*, 22 January 2003.
93. http://www.gmwatch.org/archive2.asp?ArcId=43 Elizabeth Weise,
    'Research piglets sold as food hard to find', *USA Today*, 7 February
    2003.

94. http://www.foe.org/new/releases/0503cornshow.html 'USDA sold potentially toxic corn to food and feed handler. May have violated Cargill policy and Monsanto Grower Agreement', Friends of the Earth news release, 16 May 2003.

95. http://www.gmwatch.org/archive2.asp?ArcId=43 'Tainted pigs show up in sausage at funeral', Associated Press, 3 June 2001.

96. http://www.gmwatch.org/archive2.asp?arcid=5218 Nancy Cole, 'Competition grows in the biopharming market', *Arkansas Democrat-Gazette*, 5 May 2005.

97. http://www.gmwatch.org/archive2.asp?ArcId=391 13 November 2002.

98. Fred Pearce, 'Crops "widely contaminated" by genetically modified DNA', *New Scientist*, NewScientist.com news service, 24 February 2004.

99. http://www.gefoodalert.org/library/admin/uploadedfiles/Manufacturing_Drugs.doc

100. http://www.gmwatch.org/archive2.asp?ArcId=388 'FDA orders destruction of soybeans contaminated with genetically engineered corn', Associated Press, 11 December 2002.

101. Philip Cohen, 'GM crop mishaps unite friends and foes', *New Scientist*, NewScientist.com news service, 18 November 2002.

102. http://www.gmwatch.org/archive2.asp?ArcId=394 Justin Gillis, 'Biotech Firm Mishandled Corn in Iowa', *Washington Post*, 14 November 2002.

103. http://www.gmwatch.org/archive2.asp?ArcId=406 'Anthony G. Laos appointed to board for International Food and Agriculture Development by President Bush', *College Station*, 9 September 2002.

104. http://www.gmwatch.org/archive2.asp?arcid=271 Justin Gillis, 'US will subsidize cleanup of altered corn', *Washington Post*, 26 March 2003.

105. http://www.gmwatch.org/archive2.asp?arcid=319 John Nichols, 'The Three Mile Island of biotech?', *The Nation*, 12 December 2002.

106. 'Pharmaceutical Corn Contaminates Soybean Harvest in Nebraska', Statement by Jane Rissler, Senior Scientist, the Union of Concerned Scientists, 13 November 2002.

107. http://www.gmwatch.org/archive2.asp?ArcId=394 Genetically Engineered Food Alert coalition press release, 13 November 2002.

108. http://www.gmwatch.org/archive2.asp?arcid=2621 Editorial, *Nature Biotechnology*, doi:10.1038/nbt0204-133, vol. 22, no. 2, February 2004, p. 133.

109. http://www.safe-food.org/-issue/scientists.html

110. Personal communication with Joseph Cummins, cited in Smith, *Seeds of Deception*, p. 64.

111. http://www.gmwatch.org/archive2.asp?arcid=2712 New research on survival of CaMV promoter in rat tissues, Dr Terje Traavik, February 2004.

112. *Ibid.*

113. Smith, *Seeds of Deception*, pp. 59–60.

114. *Daily Mail*, 17 July 2002.

115. Smith, *Seeds of Deception*, p. 60.

116. Ricki Lewis, 'The rise of antibiotic-resistant infections FDA page', *FDA Consumer magazine*, September 1995, cited in Smith, *Seeds of Deception*, p. 139.
117. BBC Radio Four, *Today Programme*, 26 June 2004.
118. http://www.gmwatch.org/p1temp.asp?pid=3&page=1
119. http://www.btinternet.com/~clairejr/Pusztai/puszta_1.html *GM-Free* website.
120. 1) J.A. Nordlee et al., *New England Journal of Medicine*, vol. 688, 1996. 2) A.N. Mayeno et al., *Tibtech*, vol. 12, 1994, p. 364.
121. Story of Harry Schulte, WCPO-TV 9, 11pm News, Cincinnati, Ohio, 26 February 1998, cited in Smith, *Seeds of Deception*, p. 108.
122. Sheldon Rampton and John Stauber, *Trust Us We're Experts* (Jeremy P. Tarcher/Putnam, New York, 2001), cited in Smith, *Seeds of Deception*, p. 108.
123. 1) P. Raphals, 'Does medical mystery threaten biotech?', *Science*, vol. 249, no. 619, 1990, cited in Smith, *Seeds of Deception*, p. 114. 2) E.A. Belongia and others, 'An investigation of the cause of eosinophilia-myalgia syndrome associated with tryptophan use', *New England Journal of Medicine*, 9 August 1990, cited in Smith, *Seeds of Deception*, pp. 114–15.
124. P. Raphals, 'Does medical mystery threaten biotech?', *Science*, vol. 249, no. 619, 1990, cited in Smith, *Seeds of Deception*, p. 120.
125. William Crist (an investigative reporter who has spent years studying the EMS tragedy), investigative report on L-tryptophan, found at www.biointegrity.org, cited in Smith, *Seeds of Deception*, p. 117.
126. Smith, *Seeds of Deception*, p. 125.
127. 1) P. Raphals, 'Does medical mystery threaten biotech?', *Science*, vol. 249, no. 619, 1990, cited in Smith, *Seeds of Deception*, p. 114. 2) William Crist, investigative report on L-tryptophan, found at www.biointegrity.org, cited in Smith, *Seeds of Deception*, pp. 118, 120–1.
128. 1) William Crist, investigative report on L-tryptophan, found at www.biointegrity.org, cited in Smith, *Seeds of Deception*, pp. 117–18, 121. 2) Douglas L. Archer, Deputy Director, Center for Food Safety and Applied Nutrition, FDA, Testimony before the Subcommittee on Human Resources and Intergovernmental Relations Committee on Government Operations, House of Representatives, 18 July 1991, cited in Smith, *Seeds of Deception*, pp. 121–3.
129. 1) P. Raphals, 'Does medical mystery threaten biotech?', *Science*, vol. 249, no. 619, 1990, cited in Smith, *Seeds of Deception*, p. 114. 2) E.A. Belongia et al., 'An investigation of the cause of eosinophilia-myalgia syndrome associated with tryptophan use', *New England Journal of Medicine*, 9 August 1990, cited in Smith, *Seeds of Deception*, pp. 114–15. 3) William Crist, investigative report on L-tryptophan, found at www.biointegrity.org, cited in Smith, *Seeds of Deception*, pp. 115–17.
130. Douglas L. Archer, Deputy Director, Center for Food Safety and Applied Nutrition, FDA, Testimony before the Subcommittee on Human Resources and Intergovernmental Relations Committee on Government Operations, House of Representatives, 18 July 1991, cited in Smith, *Seeds of Deception*, pp. 121–3.

131. Smith, *Seeds of Deception*, p. 119.
132. Communication with author.
133. Smith, *Seeds of Deception*, p. 75.
134. Mark Townsend, 'Why soya is a hidden destroyer', *Daily Express*, 12 March 1999, cited in Smith, *Seeds of Deception*, pp. 160–1.
135. Rick Weiss, 'Biotech food raises a crop of questions', *Washington Post*, 15 August 1999, p. A1, cited in Smith, *Seeds of Deception*, p. 163.
136. Louis J. Pribyl, 'Biotechnology Draft Document, 2/27/92', 6 March 1992, cited in Smith, *Seeds of Deception*, p. 164.
137. Michael Hansen, 'Bt crops: Inadequate testing', lecture delivered at Universidad Autonoma, Chapingo, Mexico, 2 August 2002, cited in Smith, *Seeds of Deception*, p. 177.
138. http://www.i-sis.org.uk/biopesticide&bioweapons.php
139. Bill Freese, 'A critique of the EPA's decision to re-register Bt crops and an examination of the potential allergenicity of Bt proteins', adapted from Friends of the Earth submission to the EPA, 9 December 2001, cited in Smith, *Seeds of Deception*, pp. 178–9.
140. Smith, *Seeds of Deception*, p. 179.
141. http://www.btinternet.com/~clairejr/Pusztai/puszta_1.html *GM-Free* website.
142. 1) http://www.gmwatch.org/archive2.asp?ArcId=303 'GM expert warns of cancer risk from crops', *Sunday Herald*, 8 December 2002. 2) Communication with author.
143. *Guardian*, 19 March 1998.
144. http://www.gmwatch.org/p1temp.asp?pid=3&page=1
145. http://www.mindfully.org/GE/GE4/BMA-GM-Crop-Trials20nov02.htm Submission of the British Medical Association to the Health and Community Care Committee of the Scottish Parliament, on the impact of GM Crop trials, HC/02/30/A 20nov02, 20 November 2002.
146. http://www.bio-integrity.org/ext-summary.html 'How the US FDA approved genetically engineered foods despite the deaths one had caused and the warnings of its own scientists about their unique risks', Steven M. Druker, JD, Executive Director, Alliance for Bio-Integrity.
147. http://www.gmwatch.org/p1temp.asp?pid=3&page=1
148. http://www.gmwatch.org/archive2.asp?arcid=939 George Monbiot, 'Let's do a Monsanto', *Guardian*, 10 June 2003.
149. Soil Association, 'Seeds of Doubt', p. 5.
150. Sight Savers literature, 2003. £17 (c. $25) restores sight to an adult suffering from cataracts.
151. Rowell, *Don't Worry*, p. 6.
152. http://www.gmwatch.org/archive2.asp?arcid=1368 'Genetically modified food fears seen hurting Canada', *Globe and Mail* (Canada) (via Agnet), 28 August 2003.
153. Soil Association, 'Seeds of Doubt', p. 43.
154. D. McGuire, presentation to 2002 Annual Convention of the American Corn Growers' Association, 9 March 2002, cited in Soil Association, 'Seeds of Doubt', p. 44.
155. http://www.gmwatch.org/archive2.asp?arcid=4722 Alan Guebert, 'US becomes net food importer for first time in 50 years', *Peoria Journal Star*, 7 December 2004.

156. Soil Association, 'Seeds of Doubt', p. 5.
157. *Ibid.*, p. 46.
158. *Ibid.*, p. 14.
159. Interview with Gale Lush, 27 January 2002, cited in Soil Association, 'Seeds of Doubt', p. 39.
160. http://www.terradaily.com/2004/040510162318.1rnqs61k.html 'Greenpeace paralyses Italian ports to protest genetically-modified soya', Terra.wire (Rome), 10 May 2004.
161. http://www.gmwatch.org/archive2.asp?arcid=1839 Ann Scholl and Facundo Arrizabalaga, 'Argentina: The catastrophe of GM soya', *Green Left Weekly*, 12 November 2003.
162. Paul and Steinbrecher, *Hungry Corporations*, p. 205.
163. http://www.gmwatch.org/archive2.asp?arcid=4499 Lilian Joensen and Stella Semino, 'Argentina's torrid love affair with the soybean', GRAIN, posted 8 October 2004.
164. http://www.gmwatch.org/archive2.asp?arcid=3280 'Argentina's bitter harvest', *New Scientist*, 17 April 2004.
165. *Daily Mail*, 6 May 2004.
166. http://www.gmwatch.org/archive2.asp?arcid=1839 Ann Scholl and Facundo Arrizabalaga, 'Argentina: The catastrophe of GM soya', *Green Left Weekly*, 12 November 2003.
167. http://www.gmwatch.org/archive2.asp?arcid=3268 Anastasia Stephens, 'Puncturing the GM myths', *Evening Standard*, 8 April 2004, from an interview with Dr Mae-Wan Ho, Director of the Institute of Science in Society.
168. http://www.gmwatch.org/archive2.asp?arcid=989 'Selling Food. Health. Hope: The real story behind Monsanto Corporation Part 2', a report by MASIPAG (Farmer–Scientist Partnership for Development, Inc, Philippines), 17 June 2003.
169. http://www.gmwatch.org/archive2.asp?arcid=3280 'Argentina's bitter harvest', *New Scientist*, 17 April 2004.
170. http://www.gmwatch.org/archive2.asp?arcid=1839 Ann Scholl and Facundo Arrizabalaga, 'Argentina: The catastrophe of GM soya', *Green Left Weekly*, 12 November 2003.
171. 1) http://www.gmwatch.org/archive2.asp?arcid=4820 2) http://www.gmwatch.org/archive2.asp?arcid=4192
172. *Genetix Update*, issue 26, Spring 2004, p. 5.
173. 1) http://www.gmwatch.org/archive2.asp?arcid=1566 'No-one will insure GM crops', FARM press release, 7 October 2003. 2) Kristen Philipkoski, 'Food biotech is risky business', *Wired News*, 1 December 2003.
174. http://www.gmwatch.org/archive2.asp?ArcId=371 Soil Association press release, 9 November 2002.
175. Kristen Philipkoski, 'Food biotech is risky business', *Wired News*, 1 December 2003.
176. http://www.aais.org/Viewpoint/01fall2.html 'Breaking new ground', *Viewpoint*, the American Association of Insurance Services, vol. 26, no. 2, Fall 2001.

177. Kristen Philipkoski, 'Food biotech is risky business', *Wired News*, 1 December 2003.
178. http://www.gmwatch.org/archive2.asp?ArcId=2067 Institute for Agriculture and Trade Policy press release, 6 December 2001.
179. Shivani Chaudhry, Research Foundation for Science, Technology and Ecology, New Delhi, India, cited in Five Year Freeze report, 'Feeding or Fooling the World?', pp. 30–1.
180. Anderson, *Genetic Engineering, Food, and Our Environment*, p. 72.
181. *Ibid.*, p. 80.
182. J. Madeley, 'Yours For Food', Christian Aid UK report, 1996, cited in Anderson, *Genetic Engineering, Food, and Our Environment*, p. 82.
183. Elissa Blum, 'Making biodiversity conservation profitable: A case study of the Merck/INBio Agreement', *Environment*, vol. 35, no. 4, May 1993, pp. 16–20, 38–44, cited in Anderson, *Genetic Engineering, Food, and Our Environment*, p. 82.
184. http://www.gmwatch.org/archive2.asp?arcid=497
185. http://www.gmwatch.org/archive2.asp?arcid=2902
186. 1) http://www.gmwatch.org/archive2.asp?ArcId=281 *Times* (India), 30 November 2002. 2) http://www.gmwatch.org/archive2.asp?ArcId=311 Press Trust of India News, 10 December 2002.
187. http://www.gmwatch.org/archive2.asp?ArcId=311 Press Trust of India News, 10 December 2002.
188. http://www.gmwatch.org/archive2.asp?ArcId=174 'Opposing patents on genes, proteins possible through product of nature doctrine', University of North Carolina at Chapel Hill, 2 January 2003.
189. http://www.gmwatch.org/archive2.asp?ArcId=174
190. http://www.gmwatch.org/archive2.asp?arcid=4903 Vandana Shiva, 'Sowing the seeds of dictatorship', ZNet, 14 February 2005.
191. Michael Pollan, 'Playing God in the garden', *New York Times*, 25 October 1998, cited in *GM-Free*, vol. 1, no. 1, April 1999, p. 20.
192. http://www.gmwatch.org/archive2.asp?arcid=2622
193. Five Year Freeze, 'Feeding or Fooling the World?', p. 30.
194. http://www.corporatewatch.org.uk/profiles/biotech/monsanto/monsanto5.html
195. http://www.gmwatch.org/archive2.asp?arcid=2622 An interview with RAFI, the Canada-based Rural Advancement Foundation International, and US Department of Agriculture (USDA) spokesman, Willard Phelps.
196. Paul and Steinbrecher, *Hungry Corporations*, p. 86.
197. *Ibid.*, p. 201.
198. 'Traitor Technology – damaged goods from the gene giants', RAFI news release, 29 March 1999, cited in Anderson, *Genetic Engineering, Food, and Our Environment*, p. 68.
199. Paul and Steinbrecher, *Hungry Corporations*, p. 201.
200. Anderson, *Genetic Engineering, Food, and Our Environment*, p. 68.
201. Paul and Steinbrecher, *Hungry Corporations*, p. 202.
202. http://www.gmwatch.org/archive2.asp?ArcId=174 Percy Schmeiser, 'Genetic contamination and farmers' rights', *Synthesis/Regeneration*, vol. 29, Fall 2002.

203. Soil Association, 'Seeds of Doubt', p. 47.

204. http://www.gmwatch.org/archive2.asp?arcid=3417 'Percy Schmeiser – the man that took on Monsanto', *Ecologist*, May 2004.

205. http://www.gmwatch.org/archive2.asp?ArcId=174 Percy Schmeiser, 'Genetic contamination and farmers' rights', *Synthesis/Regeneration*,vol. 29, Fall 2002.

206. http://www.gmwatch.org/archive2.asp?arcid=1372 Stephen Leahy, 'A farmer prepares for final battle against Monsanto', tierramerica.net, 25 August 2003.

207. http://www.gmwatch.org/archive2.asp?ArcId=174 Percy Schmeiser, 'Monsanto lying about 98 per cent of crop being genetically modified', *Synthesis/Regeneration*, vol. 29, Fall 2002.

208. http://www.gmwatch.org/archive2.asp?arcid=963

209. http://www.gmwatch.org/archive2.asp?ArcId=438 'Monsanto awarded $780K judgment', *Daily Journal*, djournal.com, November 2002.

210. Monsanto press release, 1998, cited in Soil Association, 'Seeds of Doubt', p. 47.

211. Peter Shinkle, 'Monsanto reaps some anger with hard line on reusing seed', CropChoice news, 12 May 2003.

212. http://www.gmwatch.org/archive2.asp?arcid=830 'Monsanto reaps some anger with hard line on reusing seed', *St Louis Post-Dispatch*, May 2003.

213. http://www.gmwatch.org/archive2.asp?arcid=3711 8 June 2004. 'We've been hearing from the BIO-devastation gathering in San Francisco that a page-size ad has been doing the rounds which is said to have been placed by Monsanto in Chiapas, Mexico. It is said this ad was repeatedly run in local newspapers in Chiapas directly in the wake of the Canadian decision. The ad is addressed to "Amigo Agricultor" (Dear Farmer Friend). In the accompanying message Monsanto reminds its friends, the farmers, that planting its seeds (or seeds carrying its patented genes) without its permission is a federal crime under the law of Mexico, and – they apparently say – can be punished by nine years in jail!'

214. http://www.gmwatch.org/archive2.asp?arcid=1254

215. http://www.gmwatch.org/archive2.asp?arcid=1254 'GM crop farmers "will end up serfs"', *Western Morning News*, 9 August 2003.

216. Soil Association, 'Seeds of Doubt', p. 22.

217. *Ibid.*, p. 39.

218. http://www.mindfully.org/GE/GE4/Heartbreak-In-The-Heartland 21jul02.htm Percy Schmeiser, and others, speaking at 'Genetically engineered seeds of controversy: Biotech bullies threaten farmer and consumer rights', the University of Texas at Austin, 10 October 2001.

219. http://www.gmwatch.org/archive2.asp?arcid=823 *Farmers Weekly* (UK), 9 May 2003.

220. http://www.btinternet.com/~nlpwessex/Documents/LordSainsbury.htm

221. http://www.gmwatch.org/archive2.asp?arcid=939 George Monbiot, 'Let's do a Monsanto', *Guardian*, June 10, 2003.

222. http://ngin.tripod.com/farming.htm

223. R. Fraley (then co-president of Monsanto's agricultural sector), *Farm Journal*, October 1996, p. 19, cited in Paul and Steinbrecher, *Hungry Corporations*, p. 24.
224. http://www.gmwatch.org/archive2.asp?arcid=4538 'World Food Day: Iraqi farmers aren't celebrating', GRAIN, 15 October 2004.
225. http://www.gmwatch.org/archive2.asp?arcid=4998
226. http://www.gmwatch.org/archive2.asp?arcid=4562
227. http://www.gmwatch.org/archive2.asp?arcid=4856 Jeremy Smith, 'Order 81', *Ecologist*, January 2005.
228. http://www.gmwatch.org/archive2.asp?arcid=4484 Stephen Leahy, 'Monsanto victory plants seed of privatisation', IPS News, 5 October 2004.
229. http://www.gmwatch.org/archive2.asp?arcid=1582 Wilkinson et al., 'Hybridization between *Brassica napus* and *B. rapa* on a national scale', *Science*, vol. 302, 2003, pp. 401–3.
230. http://www.gmwatch.org/archive2.asp?arcid=1510 Dr Peter R. Wills, Department of Physics, University of Auckland, 24 September 2003.
231. Dr John Fagan, NLP Policy statement on Genetic Engineering, 1995, pp. 110–12.
232. Alan Cooper, 'Persistence of ancient DNA in soil has implications for GE', Royal Society New Zealand news, 22 April 2003.
233. http://www.gmwatch.org/archive2.asp?arcid=2291 Gebhard and Smalla, 'Monitoring field releases of genetically modified sugar beets for persistence of transgenic plant DNA and horizontal gene transfer', *FEMS Microbiology Ecology*, 1999, vol. 28, no. 3, 1999, pp. 261–72.
234. Dr John Fagan, NLP Policy statement on Genetic Engineering, 1995, pp. 107–8.
235. Anderson, *Genetic Engineering, Food, and Our Environment*, p. 39.
236. http://www.gmwatch.org/archive2.asp?arcid=1510 Dr Peter R. Wills, Department of Physics, University of Auckland, 24 September 2003.
237. http://www.gmwatch.org/archive2.asp?arcid=1661 Peter Rosset, 'UK study finds new risks in genetically engineered crops', IPS Columnist Service, October 2003.
238. Anderson, *Genetic Engineering, Food, and Our Environment*, p. 29.
239. http://www.gmwatch.org/p2temp2.asp?aid=25&page=1&op=1
240. http://www.gmwatch.org/archive2.asp?arcid=1722
241. Anderson, *Genetic Engineering, Food, and Our Environment*, p. 24.
242. http://www.btinternet.com/~nlpwessex/Documents/gmoquote.htm
243. Erwin Chargaff, Professor Emeritus of Biochemistry, Columbia University, cited in Anderson, *Genetic Engineering, Food, and Our Environment*, p. 35.

## CHAPTER 6 THE BIOTECH LOBBY'S
## DIRTY TRICKS DEPARTMENT, PART 1

1. Monbiot, *Captive State*, p. 358.
2. http://www.spectrezine.org/environment/GMO2.htm Steve McGiffen, 'Planting lies', Spectrezine.

3. http://www.gmwatch.org/archive2.asp?arcid=1525 Michael McCarthy, 'Blair the key as decision nears over commercial GM crops in Britain', *Independent*, 25 September 2003.

4. 'HIV/AIDS in Africa', Concern literature, p. 3, 2003. £5/month would buy enough Plumpy'nut to feed a malnourished African child for a month – i.e. £60 or c. $90 a year.

5. http://www.capitaleye.org/inside.asp?ID=92 Vikki Kratz, 'Food fight', *Capital Eye*, 9 July 2003. http://www.capitaleye.org/bio-contributions.asp

6. http://www.purefood.org/Monsanto/MonBushAdmin.cfm Robert Cohen, 'Monsanto and G.W. Bush Administration: Who will own the store?', purefood.org, 21 January 2001.

7. http://www.mindfully.org/GE/GE4/Heartbreak-In-The-Heartland21jul02.htm Percy Schmeiser, and others, speaking at 'Genetically engineered seeds of controversy: Biotech bullies threaten farmer and consumer rights', the University of Texas at Austin, 10 October 2001.

8. http://www.purefood.org/Monsanto/MonBushAdmin.cfm Robert Cohen, 'Monsanto and G.W. Bush Administration: Who will own the store?', purefood.org, 21 January 2001.

9. Robert Cohen testimony before FDA panel, 2 December 1999, cited in Smith, *Seeds of Deception*, p. 148.

10. http://www.capitaleye.org/inside.asp?ID=92 Vikki Kratz, 'Food fight', *Capital Eye*, 9 July 2003. http://www.capitaleye.org/bio-lobbying.asp – according to federal lobbying reports filed with Congress.

11. http://www.capitaleye.org/inside.asp?ID=92 Vikki Kratz, 'Food fight', *Capital Eye*, 9 July 2003.

12. http://www.gmwatch.org/archive2.asp?arcid=1369 Sanjay Suri, 'Corporate giants hold government strings', Inter Press Service (IPS), 28 August 2003, referring to Friends of the Earth press release, 28 August 2003.

13. Kristen Gribben, 'Monsanto', *Capital Eye*, 9 July 2003.

14. http://www.corporatewatch.org.uk/profiles/bayer/bayer4.htm

15. Kurt Eichenwald and others, 'Biotechnology food: From the lab to a debacle', *New York Times*, 25 January 2001, cited in Smith, *Seeds of Deception*, p. 149.

16. http://www.gmwatch.org/archive2.asp?ArcId=243 Excerpt from a speech by University of Guelph-based agronomist Ann Clark, 'Genetically engineered crops: A Luddite's view'.

17. Robert Cohen, 'Monsanto and G.W. Bush Administration: Who will own the store?', purefood.org, 21 January 2001.

18. Monbiot, *Captive State*, p. 240.

19. http://www.gmwatch.org/archive2.asp?ArcId=1560 A GMWatch Special Investigation by GM Watch editor, Claire Robinson, 3 October 2003.

20. http://www.gmwatch.org/archive2.asp?ArcId=1560 1) Peter Shinkle, 'Monsanto reaps some anger with hard line on reusing seed: Agriculture giant has won millions in suits against farmers', *St Louis Post-Dispatch*, 12 May 2003. 2) Richard Thompson, 'West Tennessee grower awaits its fine in Monsanto case', *The Commercial Appeal* (US), 7 January 2003.

21. http://www.gmwatch.org/archive2.asp?ArcId=1560 'Judge rejects class action against seed producers', *New York Times*, 2 October 2003.

22. http://www.gmwatch.org/archive2.asp?ArcId=1560 http://record.wustl. edu/archive/2000/10-09-00/articles/law.html 'Law school honors six outstanding alumni', Washington University law school website, 9 October 2000.

23. http://www.gmwatch.org/archive2.asp?ArcId=1560 http://www. opensecrets.org Money in politics data.

24. http://www.gmwatch.org/archive2.asp?ArcId=1560 Rick Desloge, 'Key Thompson lawyers defect to Husch', *St Louis Business Journal*, 24 December 2001.

25. http://www.gmwatch.org/archive2.asp?arcid=952 John Ingham, 'Labour donor tells prince to keep views to himself; keep your nose out, Charles', *Express*, 16 June 2003.

26. http://www.gmwatch.org/profile1.asp?PrId=116

27. Monbiot, *Captive State*, p. 298.

28. http://www.gmwatch.org/profile1.asp?PrId=116

29. http://www.gmwatch.org/archive2.asp?arcid=1060 Andrew Rowell, 'Sinister sacking and the trail that leads to Blair and the White House', *Daily Mail*, 7 July 2003.

30. http://www.gmwatch.org/profile1.asp?PrId=116

31. *The Times*, 17 April 2002.

32. http://www.gmwatch.org/profile1.asp?PrId=116

33. http://www.gmwatch.org/archive2.asp?arcid=2900 Rob Evans and David Leigh, 'Downing St forced to reveal secret meetings', *Guardian*, 16 March 2004.

34. http://www.gmwatch.org/archive2.asp?arcid=4824 Matthew Tempest and agencies, 'Vaccine-row donor gave Labour GBP500,000', *Guardian*, 24 August 2004.

35. http://www.gmwatch.org/archive2.asp?arcid=5220

36. *Ibid.*, pp. 208–24.

37. http://www.spectrezine.org/environment/GMO2.htm Steve McGiffen, 'Planting lies', Spectrezine.

38. www.genewatch.org/Press%20Releases/pr15.htm 'Monsanto's "desperate" propaganda campaign reaches global proportions', GeneWatch UK press release, 6 September 2000, which discusses a Monsanto internal document leaked to GeneWatch UK. Monsanto's ten-page internal report, headed 'company confidential', summarises the activities of its Regulatory Affairs and Scientific Outreach teams for May and June 2000.

39. *Ibid.* Confidential report available from www.genewatch.org

40. Sarah Boseley, 'WHO "infiltrated by food industry"', *Guardian*, 9 January 2003.

41. http://www.gmwatch.org/archive2.asp?arcid=1369 Sanjay Suri, 'Corporate giants hold government strings', Inter Press Service (IPS), 28 August 2003, referring to Friends of the Earth press release, 28 August 2003.

42. List of participants, Conference of the Parties to the Convention on Biological Diversity (CBD), First Extraordinary Meeting (resumed

session), 24–28 January 2000, cited in Paul and Steinbrecher, *Hungry Corporations*, p. 156.

43. http://www.gmwatch.org/archive2.asp?arcid=4487 'BASF threatens transfer of plant research abroad', in-PharmaTechnologist.com, 6 October 2004.

44. http://www.gmwatch.org/archive2.asp?arcid=1660 1) Elizabeth Johnson, 'Monsanto puts $40m Argentine investment on hold', *Chemical News & Intelligence*, 17 October 2003. 2) 'Argentine GM policy endangers investment – Monsanto', Reuters News Service, 13 December 2000.

45. http://www.gmwatch.org/archive2.asp?arcid=1660 1) 'Monsanto and Novartis blackmail Ireland', *Corporate Europe Observer*, Corporate Europe Observatory, Holland, May 1998. 2) http://www.parliament.the-stationery-office.co.uk/pa/ld199899/ldselect/ldeucom/11/11we48.htm The UK Parliament, Memorandum by the Soil Association, reference no. 22, Affidavit to Irish High Court, section 47d, 1997 by Monsanto Europe.

46. Paul and Steinbrecher, *Hungry Corporations*, p. 162.

47. Confidential report available from www.genewatch.org, cited in Paul and Steinbrecher, *Hungry Corporations*, p. 162.

48. *Ibid.*

49. www.genewatch.org/Press%20Releases/pr15.htm 'Monsanto's "desperate" propaganda campaign reaches global proportions', GeneWatch UK press release, 6 September 2000.

50. Paul and Steinbrecher, *Hungry Corporations*, p. 162.

51. Reuters, 'Sri Lanka's GM food ban delayed indefinitely', *Times of India*, 3 September 2001, cited in Paul and Steinbrecher, *Hungry Corporations*, p. 162.

52. http://www.twnside.org.sg/title2/service117.htm 'Thai Environment Minister questions GMOs introduction through trade agreement with US', Third World Network website, 19 June 2004.

53. Paul and Steinbrecher, *Hungry Corporations*, p. 206.

54. *Ibid.*, p. 190.

55. http://www.gmwatch.org/archive2.asp?arcid=4023 'Africa – the new frontier for the GE industry', Mariam Mayet, African Centre for Biosafety, 1 July 2004.

56. http://www.gmwatch.org/archive2.asp?arcid=866 'Novel genetically engineered cotton crop to be tested in SA', SAFeAGE press release (Cape Town), 20 May 2003.

57. http://www.gmwatch.org/archive2.asp?arcid=4238 17 August 2004.

58. http://www.gmwatch.org/archive2.asp?arcid=4491 Graeme O'Neill, 'Biotech's growing in S Africa, conference hears', *Australian Biotechnology News*, 5 October 2004.

59. http://www.gmwatch.org/archive2.asp?arcid=4238 17 August 2004.

60. http://www.gmwatch.org/archive2.asp?arcid=4285 'Monsanto finds South Africa a handy laboratory and gateway to Africa', *Business Day*, 24 August 2004.

61. http://www.gmwatch.org/archive2.asp?arcid=3633

62. http://www.gmwatch.org/archive2.asp?arcid=4166 Mariam Mayet, 'African agriculture under genetic engineering onslaught', African Centre for Biosafety, 22 July 2004.

63. http://www.gmwatch.org/archive2.asp?arcid=4023 Mariam Mayet, 'Africa – the new frontier for the GE industry', African Centre for Biosafety, 1 July 2004.

64. http://www.gmwatch.org/profile1.asp?PrId=165&page=U

65. http://www.gmwatch.org/archive2.asp?arcid=3632 31 May 2004.

66. http://www.gmwatch.org/archive2.asp?arcid=4747 Konchora Guracha, 'Farmers oppose genetically modified foods bill', *The Standard* (Kenya), 20 December 2004.

67. http://www.gmwatch.org/archive2.asp?arcid=4287 'Draft biosafety bill will not protect Kenya from the risks of GMOS', press release by a coalition of farmers' groups, environmentalists and development NGOs, including Kenya Small Scale Farmers' Forum (KESSFF), Participatory Ecological Land Use Management (PELUM), Action Aid, Intermediate Technology Development Group (ITDG), ECOTERRA International, Bridge Africa, INADES, and Southern & Eastern Africa Trade Information Network Initiative (SEATINI), 2 September 2004.

68. 1) http://www.gmwatch.org/profile1.asp?PrId=165&page=U 2) http://www.gmwatch.org/archive2.asp?arcid=4166 Mariam Mayet, 'African agriculture under genetic engineering onslaught', African Centre for Biosafety, 22 July 2004.

69. http://www.gmwatch.org/profile1.asp?PrId=165&page=U

70. http://www.gmwatch.org/archive2.asp?arcid=4066 'Federal government advised on genetically modified products', *Sunday Times* (Nigeria), 4 July 2004.

71. http://www.gmwatch.org/archive2.asp?arcid=4922 Centre for Sustainable Agriculture, Andhra Pradesh, report, 'The Story of Bt Cotton in Andhra Pradesh: Erratic Processes and Results', February 2005.

72. http://www.gmwatch.org/archive2.asp?arcid=1406 Gene Campaign's comments on the official report of the Government of Andhra Pradesh on Bt cotton, 9 September 2003.

73. http://www.gmwatch.org/archive2.asp?arcid=4922 Centre for Sustainable Agriculture, Andhra Pradesh, report, 'The Story of Bt Cotton in Andhra Pradesh: Erratic Processes and Results', February 2005.

74. http://www.gmwatch.org/archive2.asp?arcid=1378 Dr Pushpa Bhargava, 'High stakes in agro research: Resisting the push', *Economic and Political Weekly*, 23 August 2003.

75. http://www.gmwatch.org/archive2.asp?ArcId=4179 Press release by P.V. Satheesh, convenor of the new study on Bt cotton by the Andhra Pradesh Coalition in Defence of Diversity (APCIDD), 30 April 2004.

76. http://www.gmwatch.org/archive2.asp?arcid=4234 Vandana Shiva, 'Regulating biotechnology', *Financial Express*, 14 August 2004.

77. http://www.gmwatch.org/archive2.asp?arcid=1427 Official report of the Government of the State of Andhra Pradesh, India, on the performance of GM Bt cotton in the season 2002.

78. http://www.gmwatch.org/archive2.asp?arcid=949 P.V. Satheesh's (of the Deccan Development Society) unpublished letter to the *Guardian*, June 2003.

79. *Ibid.*, referring to the 'Did Bt cotton save farmers in Warangal?' report from the Deccan Development Society in Andhra Pradesh, 2002/3 season.

80. http://www.gmwatch.org/archive2.asp?arcid=762 'Bt cotton farmers suffered losses, says Greenpeace', *The Hindu*, 17 April 2003, referring to the 'Performance of Bt Cotton in Karnataka', study from Greenpeace of three districts of Karnataka, April 2003.

81. http://www.gmwatch.org/archive2.asp?arcid=42

82. http://www.gmwatch.org/archive2.asp?arcid=3502 Lim Li Ching, 'Broken Promises. Will GM crops really help developing countries?, ISIS, May 2004.

83. http://www.gmwatch.org/archive2.asp?arcid=3405 'Performance of Bt Cotton in Andhra Pradesh: 2003–4', a study by Andhra Pradesh Coalition in Defense of Diversity (APCIDD), by M.A. Qayum and K. Sakkhari, May 2004.

84. http://www.gmwatch.org/archive2.asp?arcid=6310 Vandana Shiva, 'The pseudo-science of biotech lobbyists: The baseless Barfoot–Brookes claim that farmers and the environment have benefitted from GMOs', 27 February 2006. http://www.ourworldisnotforsale.org/showarticle.asp?search=1316

85. http://www.gmwatch.org/p1temp.asp?pid=42&page=1

86. http://ngin.tripod.com/pantsoftheyearaward.htm Rose De La Cruz, 'Greenpeace spends $170m to oppose biotech', *The Philippine Star*, 23 June 2002.

87. http://ngin.tripod.com/200802f.htm

88. http://ngin.tripod.com/200802f.htm Leilani M. Gallardo, 'Local farmers urged to adopt biotechnology', Asia Intelligence Wire, 26 June 2002.

89. http://www.gmwatch.org/profile1.asp?PrId=106&page=P

90. http://www.gmwatch.org/archive2.asp?arcid=329 Rhodina J. Villanueva, 'Probe OK of "Bt" corn, Bayan Muna solons urge', *TODAY Reporter* (Philippines), December 2002.

91. http://www.gmwatch.org/archive2.asp?arcid=989 'Selling Food. Health. Hope: The real story behind Monsanto Corporation Part 2', a report by MASIPAG (Farmer–Scientist Partnership for Development, Inc, Philippines), 17 June 2003.

92. http://www.gmwatch.org/archive2.asp?arcid=850 'Monsanto and US Aid-funded lobby group blocking anti-Bt corn bills', Kilusang Magbubukid ng Pilipinas (KMP) news release, 19 May 2003.

93. http://www.gmwatch.org/archive2.asp?arcid=4795 *Washington Post*, 12 January 2005.

94. http://www.gmwatch.org/archive2.asp?arcid=4795 Roberto Verzola, 'The GE debate in the Philippines: An update', 12 January 2005.

95. http://www.gmwatch.org/archive2.asp?arcid=3325 *Jakarta Post*, 17 March 2001.

96. http://www.gmwatch.org/archive2.asp?arcid=3311 Rendi A. Witular, 'Supreme Court urged to ban GMO', *Jakarta Post*, 25 March 2004.

97. http://www.gmwatch.org/archive2.asp?arcid=3311 'Genetically modified crops, a decade of failure – Indonesia', Friends of the Earth International, February 2004, part four, pp. 40–1.

98.  http://www.gmwatch.org/archive2.asp?arcid=4773 Jonathan Birchall (in New York), 'Monsanto fined $1.5m over Indonesia bribes', *Financial Times*, 6 January 2005.

99.  http://www.gmwatch.org/archive2.asp?arcid=4874 Antje Lorch, 'Monsanto fined for bribing Indonesian officials to avoid environmental studies for Bt cotton', ifrik – research on genetic engineering (Netherlands), 7 February 2005.

100. http://www.gmwatch.org/archive2.asp?arcid=4774

101. http://www.gmwatch.org/archive2.asp?arcid=3502 Lim Li Ching, 'Broken promises: Will GM crops really help developing countries?', ISIS, May 2004.

102. http://www.gmwatch.org/archive2.asp?arcid=863 Letter from Kobus Berger of Monsanto to Professor Dr Ir. Bungaran Saragih, Mec., the Ministry of Agriculture, Jakarta, 5 March 2003.

103. http://www.gmwatch.org/archive2.asp?arcid=3829 'New studies contradict FAO report and show that genetically engineered Bt cotton fails to benefit farmers', Communication from the Deccan Development Society, Andhra Pradesh Coalition in Defence of Diversity (APCIDD), and GRAIN, 16 June 2004.

104. http://www.gmwatch.org/archive2.asp?arcid=4242 'USA/Thailand: US threatened trade sanctions to block GM labels, says Thai FDA', just-food. com, 19 July 2001.

105. http://www.gmwatch.org/archive2.asp?arcid=4242 'GE papaya scandal in Thailand', Greenpeace (Thailand), 27 July 2004.

106. 1) http://www.gmwatch.org/archive2.asp?arcid=4271 2) http://www. gmwatch.org/archive2.asp?arcid=4262 Saowalak Phumyaem and Kamol Sukin, 'The GMO debate: Monsanto looks for a foothold', *The Nation* (Thailand), 26 August 2004.

107. http://www.gmwatch.org/archive2.asp?arcid=4182 'Illegal GE seeds found in packages sold by Department of Agriculture', Greenpeace, 27 July 2004.

108. http://www.gmwatch.org/archive2.asp?arcid=4295 Kultida Samabuddhi, 'Processed papaya exports rejected after GM rumours', *Bangkok Post*, 4 September 2004.

109. http://www.gmwatch.org/archive2.asp?arcid=4182 'Illegal GE seeds found in packages sold by Department of Agriculture', Greenpeace, 27 July 2004.

110. http://www.gmwatch.org/archive2.asp?arcid=4268 'Thai cabinet overturns GMO approval', Reuters, 31 August 2004.

111. Paul and Steinbrecher, *Hungry Corporations*, p. 190.

112. 'Monsanto's Bt cotton violates Thai plant quarantine laws and farmers' rights', BioThai, Alternative Agriculture Network, Foundation for Consumers, Greennet, Foundation for Thai Holistic Health press release, 26 September 1999, cited in Paul and Steinbrecher, *Hungry Corporations*, p. 217.

113. http://www.gmwatch.org/archive2.asp?arcid=1707 'GM-food producer puts poor farmers in touch with banned technology', *The Nation*, 15 November 2003.

114. http://www.gmwatch.org/archive2.asp?arcid=4377 Tul Pinkaew, 'Govt urged to "get its act together fast"', *Bangkok Post*, 22 September 2004.
115. Paul and Steinbrecher, *Hungry Corporations*, p. 91.
116. Muddassir Rizvi, 'Monsanto fiddles with Plant Protection Act', Inter Press Service, 31 August 1999, cited in Paul and Steinbrecher, *Hungry Corporations*, p. 163.
117. http://www.gmwatch.org/p1temp.asp?pid=42&page=1 'GM food ban – three years on', Environmental Foundation Ltd (Sri Lanka) press release, 25 June 2004.
118. Paul and Steinbrecher, *Hungry Corporations*, p. 162.
119. *Ibid.*, p. 203.
120. http://www.gmwatch.org/archive2.asp?arcid=3666
121. Paul and Steinbrecher, *Hungry Corporations*, p. 205.
122. http://www.gmwatch.org/archive2.asp?arcid=1660 'Argentine GM policy endangers investment – Monsanto', Reuters News Service, 13 December 2000.
123. http://www.gmwatch.org/archive2.asp?arcid=1660 Elizabeth Johnson, 'Monsanto puts $40m Argentine investment on hold', *Chemical News & Intelligence*, 17 October 2003.
124. http://www.gmwatch.org/archive2.asp?arcid=2301 Hilary Burke, 'Monsanto exits Argentina soy biz despite soy boom', Reuters, 18 January 2004.
125. http://www.gmwatch.org/archive2.asp?arcid=5007 Taos Turner, 'Argentina slams Monsanto for "attitude" on GMO royalties', Dow Jones Newswires, 17 March 2005.
126. http://www.gmwatch.org/archive2.asp?arcid=4555 'Legislation will be proposed to regulate commerce in genetically modified (GM) seeds', Business Latin America via NewsEdge Corporation. Source: Economist Intelligence Unit, 18 October 2004.
127. http://www.gmwatch.org/archive2.asp?arcid=989 'Selling Food. Health. Hope: The real story behind Monsanto Corporation Part 2', a report by MASIPAG (Farmer–Scientist Partnership for Development, Inc, Philippines), 17 June 2003.
128. Paul and Steinbrecher, *Hungry Corporations*, p. 162.
129. *Ibid.*, p. 215.
130. *Ibid.*, p. 90.
131. *Ibid.*, p. 214.
132. http://www.gmwatch.org/archive2.asp?arcid=4325
133. http://www.gmwatch.org/archive2.asp?arcid=4941 Alan Clendenning, 'Brazil OKs genetically modified crops', Associated Press, 3 March 2005.
134. http://www.gmwatch.org/archive2.asp?arcid=4961 Mario Osava, 'War over transgenics returns to the courts', IPS News, 9 March 2005.
135. *New York Times*, 25 January 2001.
136. 'Seeds become big transnational business in Iraq', CropChoice News, 28 November 2004.
137. B. Martineau, *First Fruit – The Creation of the Flavr Savr$^{TM}$ Tomato and the Birth of Biotech Food* (McGraw Hill, New York, 2000), p. 42, cited in Rowell, *Don't Worry*, p. 127.

138. The Pew Initiative on Food and Biotechnology, *GM Food Safety: Are Government Regulations Adequate?*, Washington 2002, cited in Rowell, *Don't Worry*, pp. 125–6.
139. http://www.gmwatch.org/archive2.asp?arcid=4866 Letter from Doug Gurian-Sherman to *Frontline*, vol. 22, issue 03, 29 January–11 February 2005.
140. http://www.gmwatch.org/archive2.asp?arcid=4628 D. Schubert and W. Freese, 'Safety Testing and Regulation of Genetically Engineered Foods', *Biotechnology and Genetic Engineering Reviews*, vol. 21, 16 November 2004.
141. 'Seeds become big transnational business in Iraq', CropChoice News, 28 November 2004.
142. http://www.globalcountry.org.uk/news.php?f=uk20030708p_gm.htm#PR Global Country of World Peace press release, 9 July 2003.
143. *Ibid.*
144. http://www.globalcountry.org.uk/news.php?f=uk20030708p_gm.htm#PR Letter from Dr James Maryanski, Biotechnology Coordinator, to Dr Bill Murray, Chairman of the Food Directorate, Canada, about the safety assessment of foods and food ingredients developed through new biotechnology, 23 October 1991. (FDA document #8 at www.biointegrity.org).
145. http://www.globalcountry.org.uk/news.php?f=uk20030708p_gm.htm#PR Global Country of World Peace press release, 9 July 2003.
146. http://www.globalcountry.org.uk/news.php?f=uk20030708p_gm.htm#PR The Public Health Association of Australia (PHAA) written comments to ANZFA, October 2000.
147. http://www.globalcountry.org.uk/news.php?f=uk20030708p_gm.htm#PR A report headed by Dr Masaharu Kawata, Assistant Professor in the School of Science at Nagoya University, published in the Japanese journal *Technology and Human Beings*, vol. 11, November 2000, pp. 24–33.
148. http://www.globalcountry.org.uk/news.php?f=uk20030708p_gm.htm#PR 'The Public Health Association of Australia (PHAA) analysed Monsanto's data from controlled studies on three of its GE plants (herbicide resistant maize and canola, and pesticide-producing corn) and in *all* three cases discovered several statistically significant differences in chemical composition (including amino acid profiles) between the GE organism and its non-GE counterpart. The PHAA report (October 2000) states that the differences in the amino acids cannot be attributed solely to the known products of the inserted genes and cautions that these plants may contain unexpected – and to date unidentified – new proteins that could be harmful to humans.'
149. http://www.gmwatch.org/archive2.asp?ArcId=371 'Some observations and proposals on the 2002–2003 Public Dialogue on possible commercialization of GM crops in the UK, for the Public Debate Steering Board, meeting 7 November 2002', 4 November 2002.
150. 'Meacher's GM charges rejected', BBC News Online, 22 June 2003.
151. http://www.gmwatch.org/archive2.asp?arcid=72 Paul Brown, 'More time for public say on GM crops', *Guardian*, 20 February 2003.

152. http://www.gmwatch.org/archive2.asp?arcid=939 George Monbiot, 'Let's do a Monsanto', *Guardian*, 10 June 2003.

153. http://www.gmwatch.org/archive2.asp?arcid=949 Editorial in *Nature*.

154. http://www.gmwatch.org/archive2.asp?arcid=939 George Monbiot, 'Let's do a Monsanto', *Guardian*, 10 June 2003.

155. http://www.gmwatch.org/archive2.asp?arcid=1159 Ian Sample, 'Naive, narrow and biased... Carlo Leifert explains why he resigned from the government's GM science review panel', *Guardian*, 24 July 2003.

156. *Ibid.*

157. http://www.gmwatch.org/archive2.asp?arcid=1208

158. http://www.gmwatch.org/archive2.asp?arcid=1159 Ian Sample, 'Naive, narrow and biased... Carlo Leifert explains why he resigned from the government's GM science review panel', *Guardian*, 24 July 2003. The report of the Science Review Panel 'mentions that Americans have eaten GM food for about seven years now and they haven't suffered'.

159. http://www.gmwatch.org/archive2.asp?arcid=939 George Monbiot, 'Let's do a Monsanto', *Guardian*, 10 June 2003.

160. Michael McCarthy, 'GM crops? No thanks', *Independent*, 25 September 2003.

161. http://www.gmwatch.org/archive2.asp?arcid=1531 1) Michael McCarthy, 'Blair the key as decision nears over commercial GM crops in Britain', *Independent*, 25 September 2003. 2) Michael McCarthy, 'GM crops? No thanks', *Independent*, 25 September 2003.

162. Download the UK government's 2003 Economic Review report and executive summary at: http://www.number10.gov.uk/output/Page3673.asp

163. http://www.i-sis.org.uk/ISPRUKG.php Lim Li Ching, 'Independent Science Panel rejects conclusions of GM Science Review', ISIS.

164. http://www.gmwatch.org/archive2.asp?arcid=1696 Robert Vint and Lim Li Ching, '"Cynical & dishonest science" in GM maize trials', ISIS, 11 November 2003.

165. http://www.gmwatch.org/archive2.asp?arcid=1635 Paul Brown and John Vidal, 'Two GM crops face ban for damaging wildlife', *Guardian*, 17 October 2003.

166. http://www.gmwatch.org/archive2.asp?arcid=1641

167. http://www.defra.gov.uk/news/2003/031013b.htm 'GM research reports published', DEFRA Information Bulletin, 13 October 2003.

168. http://www.gmwatch.org/archive2.asp?arcid=1623 'New research highlights dangers of modified crops', Friends of the Earth Europe press release, 14 October 2003.

169. http://www.gmwatch.org/archive2.asp?arcid=5015 Jim Giles, 'Transgenic crops take another knock', *Nature*, 21 March 2005.

170. http://www.gmwatch.org/archive2.asp?arcid=2834 'UK Government ignores science, Parliament and public concerns on GM crops. GeneWatch UK response to the Government's GM policy announcement', GeneWatch UK press release, 9 March 2004.

171. http://www.gmwatch.org/archive2.asp?arcid=2690 Geoffrey Lean, 'GM – the great betrayal', *Daily Mail*, 20 February 2004.

172.  http://www.gmwatch.org/archive2.asp?arcid=3125 John Mason, 'Beckett is blamed as Bayer bins GM plan', *Financial Times*, 30 March 2004.
173.  http://www.gmwatch.org/archive2.asp?arcid=3127 'GM Maize – major victory for grass-roots campaigners', GM-free Cymru press release, 31 March 2004.
174.  http://www.gmwatch.org/archive2.asp?arcid=3134 Pratima Desai, 'US data push European stocks down at end-quarter', Reuters, 31 March 2004.
175.  http://www.gmwatch.org/archive2.asp?arcid=228 Open letter from GM-free Cymru to Margaret Beckett, MP, 16 January 2003.
176.  http://www.gmwatch.org/archive2.asp?arcid=209 'Eco Soundings', Paul Brown, *Guardian*, 15 January 2003.
177.  http://www.gmwatch.org/archive2.asp?arcid=2622 An interview with RAFI, the Canada-based Rural Advancement Foundation International, and US Department of Agriculture (USDA) spokesman, Willard Phelps.
178.  http://www.gmwatch.org/archive2.asp?arcid=2622
179.  http://www.gmwatch.org/archive2.asp?ArcId=243 Excerpt from a speech by Ann Clark, 'Genetically engineered crops: A Luddite's view'.
180.  http://www.gmwatch.org/archive2.asp?arcid=1064 Antony Barnett and Mark Townsend, 'Anger at advisers' biotech links', *Observer*, 13 July 2003.
181.  http://www.gmwatch.org/archive2.asp?arcid=1570 George Monbiot, 'Force-fed a diet of hype', *Guardian*, 7 October 2003.
182.  A study by the Organic Farming Research Foundation, cited in Anderson, *Genetic Engineering, Food, and Our Environment*, p. 89.
183.  1) http://ngin.tripod.com/forcefeed.htm 2) http://ngin.tripod.com/271102d.htm 'It's wicked, when there is such an excess of non-GM food aid available, for GM to be forced on countries for reasons of GM politics... if there is an area where anger needs to be harnessed it is here.' UK Environment Minister Michael Meacher, speaking at a briefing of British parliamentarians, 27 November 2002.
184.  1) KMP rejoinder to Norman Borlaug's response in 'Letters to the Editor', *Independent*, to Walsh, 'America finds ready market', April 2000. 2) 'Farmers decry dumping of hazardous GMOs from relief agencies, Biogen firms', KMP press release, 14 April 2000. Both cited in Paul and Steinbrecher, *Hungry Corporations*, p. 208.
185.  Greenpeace website, 7 October 2002.
186.  http://www.gmwatch.org/archive2.asp?ArcId=439 UK Environment Minister Michael Meacher, speaking at a briefing of British parliamentarians, 27 November 2002.
187.  http://www.gmwatch.org/archive2.asp?arcid=1641 *Dow Jones International News*, January 2003.
188.  http://www.gmwatch.org/archive2.asp?arcid=865 Friends of the Earth International press release, 23 May 2003.
189.  http://www.gmwatch.org/print-archive2.asp?arcid=1425 'Meacher slams US for "grotesque misrepresentation"', Gaia Foundation press release, 12 September 2003.

190. 'Bush does aid his way, but the US is the world's stingiest donor', *Guardian*, 31 May 2003.

191. ActionAid report, 'GM Crops – Going Against the Grain', May 2003.

192. Pilger, *Hidden Agendas*, p. 3.

193. *Washington Post*, 6 June 2003.

194. George Monbiot, 'Enslaved by Free Trade', *New Scientist*, 31 May 2003.

195. http://www.gmwatch.org/archive2.asp?arcid=866 Devinder Sharma, 'US seeks to force-feed scientific apartheid to Third World', *Business Report* (South Africa), May 2003.

196. http://www.gmwatch.org/archive2.asp?arcid=2897 According to testimony made by USAID before the Committee on International Relations Subcommittee on Africa in the US House of Representatives on 11 March 2004, as of 7 March USAID had stopped all further food aid shipments to Port Sudan because the Government of Sudan had asked that US commodities be certified free of GMOs.

197. http://www.gmwatch.org/archive2.asp?arcid=4833 'Angola GMO ban to hurt food aid imports', *Business report* (South Africa), 25 January 2005.

198. http://www.gmwatch.org/archive2.asp?arcid=865

199. http://www.gmwatch.org/archive2.asp?arcid=865 Friends of the Earth International press release, 23 May 2003.

200. '*The White House For Sale*', Channel 4 (UK), 25 October 2004.

201. http://www.gmwatch.org/archive2.asp?arcid=3435

202. http://www.gmwatch.org/archive2.asp?arcid=1415 'Public bite back in GM trade war', Friends of the Earth press release, 11 September 2003.

203. http://www.gmwatch.org/archive2.asp?arcid=1369 Sanjay Suri, 'Corporate giants hold government strings', Inter Press Service (IPS), 28 August 2003, referring to Friends of the Earth press release, 28 August 2003.

204. http://www.gmwatch.org/archive2.asp?arcid=6533 'US did not win transatlantic GM trade dispute', Friends of the Earth press release, 10 May 2006.

205. http://www.gmwatch.org/archive2.asp?arcid=6292 'Leaked report: US misled the world on biotech foods "victory"', Friends of the Earth press release, 28 February 2006.

206. http://www.gmwatch.org/archive2.asp?arcid=6243 John Vidal, 'America's masterplan is to force GM food on the world: The reason the US took Europe to the WTO court was to prise open lucrative markets elsewhere', *Guardian*, 13 February 2006.

## CHAPTER 7 THE BIOTECH LOBBY'S
## DIRTY TRICKS DEPARTMENT, PART 2

1. http://www.gmwatch.org/archive2.asp?arcid=1869 Debate in House of Lords, 9 December 2003.

2. Monbiot, *Captive State*, pp. 299–301.

3. http://www.i-sis.org.uk/EngineeringLifeAndMind.php

4. Rowell, *Don't Worry*, p. 154.
5. Mark Clayton, 'Corporate cash and campus labs', *Christian Science Monitor*, 19 June 2001, cited in Paul and Steinbrecher, *Hungry Corporations*, p. 132.
6. Paul and Steinbrecher, *Hungry Corporations*, p. 101.
7. http://www.i-sis.org.uk/EngineeringLifeAndMind.php
8. Monbiot, *Captive State*, pp. 283–4.
9. http://www.gmwatch.org/archive2.asp?arcid=286
10. J.E. Bekelman, Y. Li and C.P. Gross, *Journal of the American Medical Association*, vol. 289, 2003, pp. 454–65.
11. http://www.gmwatch.org/archive2.asp?arcid=2710 George Monbiot, 'The sleaze behind our science,' *Guardian*, 24 February 2004.
12. *Ibid.*
13. http://www.gmwatch.org/archive2.asp?arcid=2899
14. Whistleblowers handbook', BBC World Service website, 15 September 2000.
15. http://www.gmwatch.org/archive2.asp?arcid=4164 *British Medical Journal*, vol. 329, no. 132, 17 July 2004.
16. http://www.gmwatch.org/archive2.asp?arcid=2710 George Monbiot, 'The sleaze behind our science', *Guardian*, 24 February 2004.
17. http://www.gmwatch.org/archive2.asp?arcid=3305 Richard Horton, 'Review: The dawn of McScience', *New York Review of Books*, vol. 51, no. 4, April 2004.
18. http://www.gmwatch.org/archive2.asp?arcid=1280 James Meek, 'Science journal accused over GM article', *Guardian*, 7 June 2002.
19. 1) http://www.gmwatch.org/archive2.asp?ArcId=540 2) http://ngin.tripod.com/110802a.htm Nature Insight feature on 'Food and the Future', *Nature*, 8 August 2002.
20. Judith Juskevich and Greg Guyer, 'Bovine Growth Hormone: Human food safety evaluation', *Science*, vol. 249, 24 August 1990, cited in Monbiot, *Captive State*, p. 234.
21. Department of Consumer Affairs, City of New York, review No. 194, cited in Monbiot, *Captive State*, p. 235.
22. http://www.gmwatch.org/archive2.asp?ArcId=85 Shanthu Shantharam, 'Bt Cotton in India – how successful is it?', AgBioView special, 21 February 2003.
23. http://www.gmwatch.org/archive2.asp?ArcId=286 Laurie Flynn and Michael Sean Gillard, *Guardian*, 1 November 1999.
24. Paul Brown, 'Printers pulp Monsanto edition of *Ecologist*', *Guardian*, 29 September 1998, cited in Smith, *Seeds of Deception*, p. 201.
25. 'Who's afraid of Monsanto? Britain's best-loved newsagents bend to history of intimidation', press release, *Ecologist*, 26 October 1998, cited in Smith, *Seeds of Deception*, p. 202.
26. http://www.gmwatch.org/archive2.asp?arcid=1431 'GM crops irrelevant for Africa', the Institute of Science in Society.
27. http://www.gmwatch.org/archive2.asp?arcid=864 'What's really going on in Makhathini', an email from Haidee Swanby of Biowatch South Africa to Jonathan Matthews of GM Watch, 20 May 2003.

28. http://www.gmwatch.org/archive2.asp?arcid=1431 Aaron deGrassi, 'Genetically modified crops and sustainable poverty alleviation in Sub-Saharan Africa: An assessment of current evidence', published by Third World Network, Africa, 24 June 2003.

29. 1) http://www.gmwatch.org/archive2.asp?arcid=5287 'Bt cotton in Makhathini, South Africa: The success story that never was', GRAIN news release (South Africa), 26 May 2005. 2) http://www.gmwatch.org/archive2.asp?arcid=5294

30. http://allafrica.com/sustainable/resources/view/00010161.pdf Aaron deGrassi, 'Genetically modified crops and sustainable poverty alleviation in Sub-Saharan Africa: An assessment of current evidence,' published by Third World Network, Africa, 24 June 2003, pp. 20–1.

31. 1) http://www.gmwatch.org/archive2.asp?arcid=864 'What's really going on in Makhathini', an email from Haidee Swanby of Biowatch South Africa to Jonathan Matthews of GM Watch, 20 May 2003. 2) http://www.safeage.org

32. http://www.gmwatch.org/archive2.asp?arcid=281 John Vidal, 'Famine in Africa: Controlling their own destiny', *Guardian*, 30 November 2002.

33. http://www.gmwatch.org/archive2.asp?ArcId=206 'GM beet research answers very few questions', Friends of the Earth press release, 15 January 2002.

34. J.D. Pidgeon, 2002, 'A novel approach to the use of genetically modified herbicide tolerant crops for environmental benefit', published in the Royal Society's *Proceedings B*, 1513, vol. 270, 15 January 2002.

35. http://www.gmwatch.org/archive2.asp?ArcId=206 'GM beet research answers very few questions', Friends of the Earth press release, 15 January 2002.

36. http://www.gmwatch.org/archive2.asp?arcid=220

37. http://www.gmwatch.org/archive2.asp?arcid=159 'Analysis and critique of Brooms Barn Research Station research report: "Economic consequences for UK farmers of growing GM herbicide tolerant sugar beet"', FARM, 20 March 2003.

38. http://www.gmwatch.org/archive2.asp?ArcId=206 Tim Radford, 'Scientists grow "bird friendly" GM sugar beet', *Guardian*, 15 January 2003.

39. http://www.gmwatch.org/archive2.asp?ArcId=206 Charles Clover, 'GM crops are "helping to save the skylark"', *Daily Telegraph*, 15 January 2003.

40. 1) http://www.gmwatch.org/archive2.asp?arcid=228 Dr Mark Avery, RSPB, 'Food for skylarks', Letters, *Independent*, 17 January 2003. 2) http://www.gmwatch.org/archive2.asp?ArcId=206 'GM beet research answers very few questions', Friends of the Earth press release, 15 January 2002.

41. http://www.gmwatch.org/archive2.asp?arcid=1623 'New research highlights dangers of modified crops', Friends of the Earth Europe press release, 14 October 2003.

42. http://www.gmwatch.org/archive2.asp?arcid=159 'Analysis and critique of Brooms Barn Research Station research report: "Economic

consequences for UK farmers of growing GM herbicide tolerant sugar beet"', FARM, 20 March 2003.

43. http://www.gmwatch.org/profile1.asp?PrId=131 Chris Lackner, 'GM crops touted to fight poverty', *National Post* (Canada), 28 June 2003.

44. http://www.gmwatch.org/profile1.asp?PrId=131 'A celebration of business innovators and ideas', *Forbes*, 5 December 2002.

45. http://www.gmwatch.org/profile1.asp?PrId=131

46. http://www.gmwatch.org/profile1.asp?PrId=131 1) *Toronto Globe and Mail*, July 2003. 2) Chris Lackner, 'GM crops touted to fight poverty', *National Post* (Canada), 28 June 2003.

47. Aaron deGrassi, 'Genetically modified crops and sustainable poverty alleviation in Sub-Saharan Africa: An assessment of current evidence', published by Third World Network, Africa, 24 June 2003.

48. http://www.gmwatch.org/profile1.asp?PrId=131 Aaron deGrassi, 'Genetically modified crops and sustainable poverty alleviation in Sub-Saharan Africa: An assessment of current evidence', published by Third World Network, Africa, 24 June 2003.

49. http://www.gmwatch.org/profile1.asp?PrId=131 Lynn J. Cook, 'Millions served: Florence Wambugu feeds her country with food others have the luxury to avoid', *Forbes*, 23 December 2002.

50. http://www.gmwatch.org/profile1.asp?PrId=131

51. *Ibid.*

52. Smith, *Seeds of Deception*, p. 264.

53. http://www.gmwatch.org/archive2.asp?arcid=3976 'Dr Mae-Wan Ho explains how the Food Standards Agency appears to be selectively promoting and suppressing research results in projects it funds', *Science in Society*, issue 22, Summer 2004.

54. http://www.gmwatch.org/archive2.asp?arcid=5266 Geoffrey Lean, 'Rats fed GM corn due for sale in Britain developed abnormalities in blood and kidneys', *Independent on Sunday*, 22 May 2005.

55. http://www.gmwatch.org/archive2.asp?arcid=5308 'Comments on the "Pusztai Report,"' Dr Brian John, GM-free Cymru, 31 May 2005.

56. http://www.gmwatch.org/archive2.asp?arcid=5266 Geoffrey Lean, 'Rats fed GM corn due for sale in Britain developed abnormalities in blood and kidneys', *Independent on Sunday*, 22 May 2005.

57. http://www.gmwatch.org/archive2.asp?arcid=3589 'Monsanto defies German government on risk study as EU Commission prepares to approve GM maize', Greenpeace Germany, 18 May 2004.

58. http://www.gmwatch.org/print-archive2.asp?arcid=5270 'GM maize conspiracy revealed', GM-free Cymru press release, 24 May 2005.

59. http://www.gmwatch.org/archive2.asp?arcid=5306 'Monsanto agrees to release of feeding study evaluations', GM-free Cymru press release, 31 May 2005.

60. http://www.gmwatch.org/archive2.asp?arcid=5308 'Comments on the "Pusztai Report"', Dr Brian John, GM-free Cymru, 31 May 2005.

61. http://www.gmwatch.org/archive2.asp?arcid=5356 'Court orders Monsanto to make scandal report public', Environmental Media Services, source: Greenpeace International, 10 June 2005.

62. http://www.gmwatch.org/archive2.asp?arcid=5415 'Commission reaction on Council votes on safeguards and GM maize MON863', official statement from EU Commission, 24 June 2005.

63. http://www.gmwatch.org/archive2.asp?arcid=5415

64. http://www.gmwatch.org/archive2.asp?arcid=1060 Andrew Rowell, 'Sinister sacking and the trail that leads to Blair and the White House', *Daily Mail*, 7 July 2003.

65. 'World renowned scientist lost his job when he warned about GE foods', Physicians and Scientists for Responsible Application of Science and Technology (PSRAST) http://www.psrast.org/pusztai.htm, cited in Smith, *Seeds of Deception*, p. 43. 'The Rowett Institute ... relies heavily on the profit of its commercial subsidiary, Rowett Research Services. This entity contracts with biotech, pharmaceutical, and other companies for research contracts, the proceeds of which help fund the Institute. Thus, the Rowett is "dependent on the industry for its existence."'

66. *GM-Free*, vol. 1, no. 1, April 1999, p. 4.

67. Smith, *Seeds of Deception*, p. 21.

68. *Ibid.*, pp. 22–3.

69. 1) http://www.gmwatch.org/archive2.asp?arcid=1060 Andrew Rowell, 'Sinister sacking and the trail that leads to Blair and the White House', *Daily Mail*, 7 July 2003. 2) Rowell, *Don't Worry*, pp. 89–91.

70. http://www.btinternet.com/~clairejr/Pusztai/puszta_4.html *GM-Free* website.

71. http://www.gmwatch.org/archive2.asp?arcid=1060 Andrew Rowell, 'Sinister sacking and the trail that leads to Blair and the White House', *Daily Mail*, 7 July 2003.

72. Marie Woolf, 'People distrust Government on GM foods', *Sunday Independent*, 23 May 1999, cited in Smith, *Seeds of Deception*, p. 30.

73. http://www.gmwatch.org/archive2.asp?arcid=1060 Andrew Rowell, 'Sinister sacking and the trail that leads to Blair and the White House', *Daily Mail*, 7 July 2003.

74. Gwynne Dyer, 'Frankenstein foods', *Globe and Mail*, 20 February 1999, cited in Smith, *Seeds of Deception*, p. 27.

75. http://www.gmwatch.org/archive2.asp?arcid=1060 Andrew Rowell, 'Sinister sacking and the trail that leads to Blair and the White House', *Daily Mail*, 7 July 2003.

76. Smith, *Seeds of Deception*, p. 33.

77. BBC Radio Four, *Seeds of Trouble*, 7 January 2003, cited in Rowell, *Don't Worry*, p. 152.

78. E. Press and J. Washburn, 'The kept university', *Atlantic Monthly*, vol. 285, no. 3, 2000, pp. 39–54, cited in Rowell, *Don't Worry*, p. 154.

79. Rowell, *Green Backlash – Global Subversion of the Environment Movement* (Routledge, 1996), cited in Rowell, *Don't Worry*, p. 154.

80. Rowell, *Don't Worry*, p. 155.

81. *Ibid.*, pp. 156–60.

82. *Ibid.*, p. 164.

83. http://www.gmwatch.org/p2temp2.asp?aid=55&page=1&op=2

84. A. Suarez et al., 'Correspondence', *Nature*, vol. 417, 27 June 2002, p. 897, cited in Rowell, *Don't Worry*, p. 164.

85. http://www.gmwatch.org/archive2.asp?arcid=1877 Rex Dalton, 'Berkeley accused of biotech bias as ecologist is denied tenure', *Nature*, vol. 426, 11 December 2003.
86. *Nature*, vol. 430, 5 August 2004, p. 598.
87. http://www.gmwatch.org/archive2.asp?arcid=4200 Goldie Blumenstyk, 'Peer reviewers give thumbs down to Berkeley–Novartis deal', *Chronicles of Higher Education*, 30 July 2004.
88. http://www.gmwatch.org/archive2.asp?arcid=5272 Richard Brenneman, 'Professor Ignacio Chapela wins bitter UC tenure fight', berkeleydaily. org, 24 May 2005.
89. http://www.gmwatch.org/archive2.asp?ArcId=288 Jonathan Matthews, 'The fake parade', *Freezerbox*, 3 December 2002.
90. http://www.gmwatch.org/archive2.asp?arcid=1159 Ian Sample, 'Naive, narrow and biased... Carlo Leifert explains why he resigned from the government's GM science review panel', *Guardian*, 24 July 2003.
91. http://www.gmwatch.org/archive2.asp?arcid=1162 'Deplorable attack on GM critic', Friends of the Earth press release, 25 July 2003.
92. http://www.gmwatch.org/p2temp2.asp?aid=50&page=1&op=1 1) 'Greenpeace wins damages over Professor's "unfounded" allegations', *Education Guardian*, 8 October 2001. 2) *Ecologist*, vol. 31, no. 10, p. 11. 3) *Private Eye*, 1040, p. 4.
93. http://ngin.tripod.com/deceit5.html Zac Goldsmith, Editor, the *Ecologist*.
94. http://www.gmwatch.org/archive2.asp?arcid=4578
95. http://www.gmwatch.org/p2temp2.asp?aid=50&page=1&op=1
96. http://www.freezerbox.com/archive/2001/04/biotech/
97. Pilger, *Hidden Agendas*, pp. 512–14.
98. *Ibid.*, p. 531.
99. *Ibid.*, p. 478.
100. Ramsey Clark, *The Fire This Time: US War Crimes In The Gulf* (Thunder's Mouth Press, 1994), p. 147, cited in Edwards, *Free to be Human*, p. 30.
101. Pilger, *Hidden Agendas*, p. 467.
102. *Ibid.*, p. 470.
103. *Ibid.*, p. 529.
104. *Ibid.*, p. 531.
105. *Ibid.*, p. 540.
106. *Ibid.*, pp. 540–1.
107. *Ibid.*, p. 496. Disclosed by Richard Norton-Taylor of the *Guardian* in 1991.
108. Pilger, *Hidden Agendas*, p. 496.
109. *Observer*, 18 August 1985, cited in Pilger, *Hidden Agendas*, p. 496.
110. Gore Vidal, *United States* (Random House, 1992), p. 1031, cited in Edwards, *Free to be Human*, p. 196.
111. 'Growth hormones would endanger milk', Op-ed article, *Los Angeles Times*, 27 July 1989, cited in Smith, *Seeds of Deception*, p. 194.
112. Rampton and Stauber, *Trust Us We're Experts*, p. 164, cited in Smith, *Seeds of Deception*, pp. 195–7.
113. Smith, *Seeds of Deception*, p. 6.

114. US newspapers present biased view of biotech', organicconsumers.org news release, 29 April 2002, cited in Smith, *Seeds of Deception*, p. 197.
115. Smith, *Seeds of Deception*, pp. 214–17.
116. Personal communication with Steven M. Druker, cited in Smith, *Seeds of Deception*, p. 216.
117. http://www.gmwatch.org/p1temp.asp?pid=33&page=1
118. http://www.gmwatch.org/archive2.asp?arcid=42 1) Letter to BBC World Service from Rod Harbinson, 8 February 2003. 2) 'Unfortunately for Monsanto, Greenpeace-India sent its own researchers to check up on how the data had been compiled and, amongst much else, the researchers collected testimonies from farmers who said that they had been advised by the company to inflate their real yield figures! Many other irregularities were also found.'
119. 'High yield from India's GM crops', BBC News Online, 7 February 2003.
120. Shaoni Bhattacharya, 'GM crops boost yields more in poor countries', *New Scientist*, NewScientist.com news service, 6 February 2003.
121. http://www.gmwatch.org/archive2.asp?arcid=62 Charlie Kronick, Chief Policy Advisor at Greenpeace UK, Letters, *The Times*, 7 February 2003.
122. http://www.gmwatch.org/archive2.asp?arcid=42 Mark Henderson, 'Farmers reap benefit of GM cotton crops', *The Times*, 7 February 2003.
123. http://www.gmwatch.org/archive2.asp?arcid=947 Mark Henderson, 'Who cares what "the people" think of GM foods?', *The Times*, 13 June 2003.
124. http://www.organicconsumers.org/corn/industrylies041502.cfm Mark Henderson, 'Attack on safety of GM crops was unfounded', *The Times*, 5 April 2002.
125. http://www.gmwatch.org/archive2.asp?arcid=348 Mark Henderson, 'Modified crops "help man and wildlife"', *The Times*, 25 August 1998.
126. Mark Henderson, 'New GM rice could transform the fight against famine', *The Times*, 26 November 2002.
127. http://www.btinternet.com/~nlpwessex/Documents/monsanto MASpossibilities.htm
128. http://www.gmwatch.org/archive2.asp?arcid=963 Pallab Ghosh, BBC, June 2003.
129. Asha Krishnakumar, 'A lesson from the field', *Frontline*, vol. 20, issue 11, 24 May–6 June 2003.
130. *Ibid.*
131. http://www.gmwatch.org/archive2.asp?arcid=963 Letter from Research Foundation for Science, Technology and Ecology (RFSTE), New Delhi, to the BBC, 20 June 2003.
132. http://www.gmwatch.org/archive2.asp?ArcId=13 BMA press release, 'BMA on GM crops: Clarification', 31 January 2003.
133. http://www.gmwatch.org/profile1.asp?PrId=203&page=G
134. http://www.gmwatch.org/archive2.asp?arcid=1708 Dick Taverne, 'Thunderer: When crops burn, the truth goes up in smoke', *The Times*, 18 November 2003.

135. http://www.gmwatch.org/archive2.asp?arcid=1708
136. http://www.gmwatch.org/archive2.asp?arcid=3427 Dick Taverne, 'The costly fraud that is organic food', *Guardian*, 6 May 2004.
137. http://www.gmwatch.org/archive2.asp?arcid=1654
138. http://www.gmwatch.org/archive2.asp?arcid=228
139. http://www.gmwatch.org/archive2.asp?arcid=2480 'GM corn approval "a disaster for democracy"', MEP news release from the office of the Green MEPs, 28 January 2004.
140. http://www.gmwatch.org/archive2.asp?arcid=2890 'Bogus comparison in GM maize trial', ISIS report, 12 March 2004.
141. http://www.gmwatch.org/archive2.asp?arcid=1845
142. http://www.gmwatch.org/profile1.asp?PrId=80&page=M
143. http://www.gmwatch.org/archive2.asp?arcid=4137 1) 'Readers consider the source, but media don't always give it; news articles often silent on scientists' and groups' funding & biases', CSPI Newsroom, 7 July 2004. 2) 'Report faults scientific journals on financial disclosure, CSPI says authors fail to disclose financial conflicts of interest; journals fail to enforce disclosure policies', CSPI Newsroom, 12 July 2004.
144. BBC Radio Four, *Today Programme*, 25 November 2003.
145. Editorial, 'Less spin, more science', *Sunday Independent*, 23 May 1999, cited in Smith, *Seeds of Deception*, p. 31.
146. http://www.gmwatch.org/archive2.asp?arcid=801 Don Westfall, in an interview with the *Toronto Star*.
147. http://www.gmwatch.org/archive2.asp?arcid=3417 'Percy Schmeiser – the man that took on Monsanto', *Ecologist*, May 2004.
148. http://www.gmwatch.org/archive2.asp?ArcId=591
149. F. Pearce, 'UN is slipping modified food into aid', *New Scientist* (London), 19 September 2002, cited in Rowell, *Don't Worry*, p. 7.
150. Emad Mekay, 'US WTO dispute could bend poor nations to GMOs', IPS, 14 May 2002.
151. Rowell, *Don't Worry*, pp. 149–50.
152. 'A bridge too far', Greenpeace news, from website, 18 August 2003.
153. http://www.gmwatch.org/archive2.asp?ArcId=171 Melissa Marino, 'GM ruling sparks fears', *The Age*, 5 January 2003.
154. http://www.i-sis.org.uk/TCCST.php 'Transgenic contamination of certified seed stocks', ISIS website.
155. Soil Association report, 'Seeds of Doubt', p. 43.
156. http://www.gmwatch.org/archive2.asp?arcid=1623 'New research highlights dangers of modified crops', Friends of the Earth Europe press release, 14 October 2003.
157. ensnews.com, 16 December 2002.
158. http://www.gmwatch.org/archive2.asp?arcid=2885 Mario Osava, 'Agriculture-Brazil: Transgenic soy found "guilty" by people's court', IPS, 11 March 2004.
159. http://www.gmwatch.org/archive2.asp?arcid=4138
160. http://www.gmwatch.org/archive2.asp?ArcId=176 *Independent*, 6 January 2003.
161. Monbiot, *Captive State*, p. 234.
162. *Ibid.*, p. 251.

163. Shiv Chopra and others, rBST (Nutrilac) 'Gaps Analysis' report by rBST Internal Review Team, Health Protection Branch, Health Canada, Ottawa, Canada, 21 April 1998, cited in Smith, *Seeds of Deception*, p. 88.

164. Samuel Epstein and Pete Hardin, 'Confidential Monsanto research files dispute many bGH safety claims', *The Milkweed*, January 1990, cited in Smith, *Seeds of Deception*, p. 88.

165. R. Torrisi and others, 'Time course of fenretinide-induced modulation of circulating insulin-like growth factor (IGF)-i, IGF-II and IGFBP-3 in bladder cancer chemo-prevention trial', *International Journal of Cancer*, vol. 87, no. 4, August 2000, pp. 601–5, cited in Smith, *Seeds of Deception*, p. 96.

166. Smith, *Seeds of Deception*, p. 98.

167. http://www.corporatewatch.org.uk/profiles/biotech/monsanto/monsanto5.html

168. Monbiot, *Captive State*, p. 234.

169. http://www.btinternet.com/~nlpwessex/Documents/gmoquote.htm – Dr Samuel Epstein.

170. *Rachel's Environment and Health Weekly*, No. 381, 17 March 1994, cited in Monbiot, *Captive State*, p. 234.

171. Office of Consumer Affairs, US FDA, 1997, *Backgrounder: Bovine Growth Hormone or Bovine Somatotropin*, cited in Monbiot, *Captive State*, p. 234.

172. Richard Burroughs, quoted in Jeff Kamen, 'Formula for disaster: An investigative report on genetically engineered Bovine Growth Hormone in milk and consequences for you health', *Penthouse Magazine*, March 1999, cited in Monbiot, *Captive State*, p. 238.

173. Robert Cohen, *Milk, the Deadly Poison* (Argus Publishing, Englewood Cliffs, N.J., 1998), cited in Smith, *Seeds of Deception*, p. 83.

174. Shiv Chopra and others, rBST (Nutrilac) 'Gaps Analysis' report by rBST Internal Review Team, Health Protection Branch, Health Canada, Ottawa, Canada, 21 April 1998, cited in Smith, *Seeds of Deception*, p. 84.

175. Monbiot, *Captive State*, pp. 234–7.

176. *Guardian*, 20 February 1999; Monbiot, *Captive State*, p. 236.

177. Judith Juskevich and Greg Guyer, 'Bovine Growth Hormone: Human food safety evaluation', *Science*, vol. 249, 24 August 1990, cited in Monbiot, *Captive State*, p. 234.

178. http://www.btinternet.com/~nlpwessex/Documents/gmoquote.htm Derrick Jensen, 'An epidemic of deception: Why we can't trust the cancer establishment', an interview with Dr Samuel Epstein, *The Sun* (US), March 2000.

179. Monbiot, *Captive State*, p. 251.

180. 'Milk, rBGH, and cancer', *Rachel's Environment and Health Weekly*, no. 593, 9 April 1998, cited in Smith, *Seeds of Deception*, p. 187.

181. 'Monsanto's dirty tricks', *Grassroots*, Pure Food Campaign, October 1995, p. 13, cited in Anderson, *Genetic Engineering, Food, and Our Environment*, p. 107.

182. Smith, *Seeds of Deception*, pp. 186–93.

183. http://www.gmwatch.org/archive2.asp?ArcId=68 Email from Steven Wilson to GM Watch, and others, 18 February 2003.

184. *Observer*, 20 June 1999, cited in Monbiot, *Captive State*, pp. 246–7.

185. Monbiot, *Captive State*, pp. 246–7.

186. http://www.gmwatch.org/archive2.asp?ArcId=480

187. http://www.gmwatch.org/archive2.asp?ArcId=368

188. Smith, *Seeds of Deception*, p. 218.

189. http://www.gmwatch.org/archive2.asp?ArcId=368

190. William K. Jaeger, 'Economic issues and Oregon Ballot Measure 27: Labeling of genetically modified foods', Oregon State University Extension Service, October 2002, cited in Smith, *Seeds of Deception*, p. 219.

191. http://www.gmwatch.org/archive2.asp?ArcId=493 'PR expert warns gene giants on no-labeling stance', *PR Week* (US), 7 October 2002.

192. http://www.gmwatch.org/archive2.asp?ArcId=501 Alliance for Bio-Integrity exposes misrepresentations in FDA's letter to Governor Kitzhaber, in a letter to the Governor, 10 October 2002.

193. http://www.mindfully.org/GE/GE4/Heartbreak-In-The-Heartland 21jul02.htm Percy Schmeiser, and others, speaking at 'Genetically engineered seeds of controversy: Biotech bullies threaten farmer and consumer rights', the University of Texas at Austin, 10 October 2001.

194. http://www.gmwatch.org/archive2.asp?arcid=2905 Sarah Sabaratnam, 'Standing up for the truth about GM food', *New Straits Times* (Malaysia), 16 March 2004.

195. Ben Lilliston and Ronnie Cummins, 'Organic vs. "organic": The corruption of a label', *Ecologist*, vol. 28, no. 4, July/August 1998, p. 197, cited in Anderson, *Genetic Engineering, Food, and Our Environment*, p. 100.

196. http://www.gmwatch.org/archive2.asp?ArcId=288 Jonathan Matthews, 'The fake parade', *Freezerbox*, 3 December 2002.

197. http://www.gmwatch.org/archive2.asp?arcid=117 Jonathan Matthews, 'Biotech's hall of mirrors', *GeneWatch*, vol. 16 no. 1, January–February 2003.

198. *Ibid.*

199. J. Byrne, 'Protecting your assets: An inside look at the perils and power of the internet', a presentation to the Ragan Communications Strategic Public Relations Conference, V-Influence, 11 December 2001, cited in Rowell, *Don't Worry*, p. 160.

200. http://www.gmwatch.org/p2temp2.asp?aid=58&page=1&op=1 George Monbiot, *Guardian*, 29 May 2002.

201. http://www.gmwatch.org/archive2.asp?arcid=276

202. http://www.gmwatch.org/archive2.asp?arcid=117 Jonathan Matthews, 'Biotech's hall of mirrors', *GeneWatch*, vol. 16 no. 1, January–February 2003.

203. http://www.gmwatch.org/archive2.asp?arcid=1721 Jane Garrison, 'Big business interests are pushed in a program to teach educators about "agriculture"... their agriculture', *Conscious Choice*, November 2003.

204. Paul and Steinbrecher, *Hungry Corporations*, pp. 53–4.

205. Rob Edwards, 'Fury at pro-GM school magazines', *Sunday Herald*, 15 April 2001.
206. http://www.gmwatch.org/profile1.asp?PrId=111
207. http://www.gmwatch.org/profile1.asp?PrId=111 Review of the play, 'Sweet as you are', by Dr Jeremy Bartlett.

## 8 SETBACKS FOR THE BIOTECH LOBBY

1. Smith, *Seeds of Deception*, p. 1. Arthur Andersen Consulting Group at a biotech industry conference in January 1999, describing how it would help Monsanto achieve its plan of controlling the world's food supply.
2. http://www.gmwatch.org/archive2.asp?arcid=2566 Devinder Sharma, 'Dear Editor', Letters, *Evening Standard*, 5 February 2004.
3. Soil Association report, 'Seeds of Doubt, North American Farmers' Experiences of GM Crops', 2002, p. 45.
4. http://www.gmwatch.org/archive2.asp?arcid=1704 Randy Fabi, 'Green groups sue USDA to stop bio-pharm planting', Reuters, 13 November 2003.
5. http://www.gmwatch.org/archive2.asp?arcid=3125
6. http://www.gmwatch.org/archive2.asp?arcid=3344
7. http://www.gmwatch.org/archive2.asp?arcid=3612
8. http://www.gmwatch.org/archive2.asp?ArcId=3466 'Biotechnology company to close GM canola program', ABC News, 12 May 2004.
9. http://www.gmwatch.org/archive2.asp?arcid=4680 Hannelore Crolly, 'Syngenta halts genetic engineering projects in Europe', *Die Welt*, 29 November 2004. http://www.gmwatch.org/archive2.asp?arcid=5162 Email from GeneWatch UK, 26 April 2005.
10. http://www.gmwatch.org/archive2.asp?arcid=4770 Dr Brian John, 'People power and the politics of the mad-house', 5 January 2005.
11. http://www.gmwatch.org/archive2.asp?arcid=4680 Hannelore Crolly, 'Syngenta halts genetic engineering projects in Europe', *Die Welt*, 29 November 2004.
12. *Financial Times*, 1 July 2004.
13. John Vidal, 'Eco sounding', *Guardian*, 28 April 2004.
14. http://www.gmwatch.org/archive2.asp?arcid=2682 Marie Woolf, '"GM-free" rebellion grows as ministers give crops backing', *Independent*, 20 February 2004.
15. http://www.gmwatch.org/archive2.asp?arcid=949
16. http://www.gmofree-europe.org
17. http://www.gmofree-europe.org
18. http://www.gmwatch.org/archive2.asp?arcid=4801 Roberto Pinton, 'Italy 2004 – Review of the year', 16 January 2005.
19. http://www.gmwatch.org/archive2.asp?ArcId=536
20. http://www.gmwatch.org/archive2.asp?arcid=3285
21. http://www.gmwatch.org/archive2.asp?arcid=1659 Reese Ewing, 'Brazil battle over biotech soy threatens top export', Reuters, 27 October 2003.
22. http://www.gmwatch.org/archive2.asp?arcid=4156

23. http://www.gmwatch.org/archive2.asp?arcid=4614 Isabelle Southcott, 'Powell River honoured for GE free zone', *The Powell River Peak*, 9 November 2004.
24. Soil Association report, 'Seeds of Doubt', p. 45.
25. forbes.com from Reuters, 21 October 2003.
26. http://www.gmwatch.org/archive2.asp?arcid=4725 'Zambia: Government drafts biosafety legislation, Reuters, 14 December 2004.
27. http://www.gmfreeireland.org 1,000 Irish GMO-free zones declared on 22 April, Earth Day 2005.
28. afxnews.com
29. http://www.gmwatch.org/print-archive2.asp?arcid=1207 'The Confédération Paysanne resumes its anti-GM actions', *Le Monde* (Paris), 24 July 2003.
30. http://www.gmwatch.org/archive2.asp?ArcId=1551 *SchNEWS* 425, 3 October 2003.
31. http://www.gmwatch.org/archive2.asp?arcid=1418 P.T. Bopanna, 'Protesters attack Monsanto greenhouse in southern India', Associated Press, 11 September 2003.
32. http://www.gmwatch.org/archive2.asp?arcid=1418 Habib Beary, 'Indian farmers target Monsanto', BBC correspondent in Bangalore, 11 September 2003.
33. 'Brazil activists target Monsanto', BBC News Online, 3 June 2003.
34. http://www.gmwatch.org/archive2.asp?arcid=2297 Matt Nippert, 'Anti-GM troops set for action', *New Zealand Herald*, 18 January 2004.
35. http://www.gmwatch.org/archive2.asp?arcid=1794 NO! GMO Campaign Japan press information, 1 December 2003.
36. http://www.gmwatch.org/archive2.asp?arcid=2896
37. http://www.gmwatch.org/archive2.asp?arcid=2784 Paul Jacobs, 'Humboldt activists follow Mendocino example', 4 March 2004.
38. http://www.gmwatch.org/archive2.asp?arcid=4772 Luke Anderson, '2004 report – Grassroots actions in California and Hawaii', 6 January 2005.
39. http://www.gmwatch.org/archive2.asp?arcid=4635
40. http://www.gmwatch.org/archive2.asp?arcid=5111 Brian Tokar, 'Agribusiness targets state legislators to pre-empt local laws on seeds', 15 April 2005.
41. http://www.gmwatch.org/archive2.asp?ArcId=556 ISIS condemns Prime Minister's Scoping Note, 25 October 2002.
42. http://www.gmwatch.org/archive2.asp?arcid=982 David P. Hamilton, 'Biotech industry could turn profitable this decade', *Dow Jones Newswires/ Wall Street Journal*, 12 June 2003.
43. http://www.gmwatch.org/archive2.asp?arcid=5353 Paul Elias, 'Biotechnology loses billions a year', Associated Press, 31 May 2005.
44. http://www.gmwatch.org/archive2.asp?arcid=982 David P. Hamilton, 'Biotech industry could turn profitable this decade', *Dow Jones Newswires/ Wall Street Journal*, 12 June 2003.
45. Ernst & Young's 2003 German Biotechnology Report and the European Biotechnology Report.

46. http://www.gmwatch.org/archive2.asp?arcid=822 Joyce Tait, director of Edinburgh University's Centre for Social and Economic Research on Innovation in Genomics (Innogen), *Nature Biotechnology*, vol. 21, no. 5, May 2003, pp. 468–69.

47. http://www.gmwatch.org/archive2.asp?ArcId=556 ISIS condemns Prime Minister's Scoping Note, 25 October 2002.

48. http://www.corporatewatch.org.uk/genetics/monsanto.htm# pharmacia

49. http://www.gmwatch.org/archive2.asp?arcid=1641 *New York Times*, 16 October 2003.

50. http://www.gmwatch.org/archive2.asp?ArcId=556 ISIS condemns Prime Minister's Scoping Note, 25 October 2002.

51. *Financial Times*, 1 July 2004.

52. http://www.gmwatch.org/archive2.asp?ArcId=557 'Biotech debacle in four parts', Mae-Wan Ho and Lim Li Ching, ISIS special briefing paper for Prime Minister's Strategy Unit, 24 October 2002.

53. http://www.gmwatch.org/archive2.asp?arcid=1362 Jason Warick, 'Most farmers say no to growing GM wheat', *StarPhoenix* (Canada), 9 August 2003.

54. 'Wheat board mulls legal action', Reuters News Agency, 19 June 2003.

55. 1) http://www.gmwatch.org/archive2.asp?ArcId=557 'Biotech debacle in four parts', Mae-Wan Ho and Lim Li Ching, ISIS special briefing paper for Prime Minister's Strategy Unit, 24 October 2002. 2) Summary of report available from: http://www.gmwatch.org/archive2.asp?ArcId=502

56. http://www.gmwatch.org/archive2.asp?ArcId=135 'Biotech wheat may cut US exports in half – study', Reuters (US), 13 March 2003.

57. http://www.gmwatch.org/archive2.asp?ArcId=547 Rank Hovis, in the North Dakota Wheat Commission's newsletter, July–August 2002.

58. http://www.gmwatch.org/archive2.asp?arcid=2373 'Africa: Dumping ground for rejected GM wheat', African Centre for Biosafety press release, 21 January 2004.

59. http://www.i-sis.org.uk/EngineeringLifeAndMind.php

60. http://www.gmwatch.org/archive2.asp?arcid=2373 'Africa: Dumping ground for rejected GM wheat', African Centre for Biosafety press release, 21 January 2004.

61. http://www.gmwatch.org/archive2.asp?ArcId=3443 Andrew Pollack, 'Monsanto shelves plan for modified wheat', *New York Times*, 11 May 2004.

62. http://www.gmwatch.org/archive2.asp?arcid=2439 'Biotech wheat pits farmer vs. farmer', *Chicago Tribune*, 25 January 2004.

63. http://www.gmwatch.org/archive2.asp?arcid=4240 Colin Perkel, 'Monsanto ripped over wheat experiments', cnews Canada, 17 August 2004.

64. http://www.gmwatch.org/archive2.asp?arcid=3465 Sue Mayer, 'Is this the end for GM food?', *Guardian*, 11 May 2004.

65. Paul and Steinbrecher, *Hungry Corporations*, pp. 203, 213.

66. Report from Latin American Meeting on Food Aid and GMOs, Quito, Ecuador, 6–9 August 2001, cited in Paul and Steinbrecher, *Hungry Corporations*, p. 209.
67. Paul and Steinbrecher, *Hungry Corporations*, p. 173.
68. Iza Kruszewska, 'Corporate influence in Central and Eastern Europe in the field of agricultural biotechnology', report, June 2001, cited in Paul and Steinbrecher, *Hungry Corporations*, p. 174.
69. Paul and Steinbrecher, *Hungry Corporations*, p. 180.

## CHAPTER 9 A MORE CONSTRUCTIVE WAY FORWARD

1. http://www.gmwatch.org/archive2.asp?arcid=4226 Mr Hodson QC, acting on behalf of the Life Sciences Network at the New Zealand Royal Commission on Genetic Modification, 8 February 2001.
2. Paul and Steinbrecher, *Hungry Corporations*, p. 38.
3. http://www.gmwatch.org/archive2.asp?arcid=3340 Richard Manning, 'Super organics', wired.com.
4. http://www.btinternet.com/~nlpwessex/Documents/GMdebatesolution.htm 'Wheat future in bio-tech not GM', *Farmers Weekly*, 25 February 2000 (Arable Focus Supplement).
5. http://www.btinternet.com/~nlpwessex/Documents/monsanto MASpossibilities.htm
6. http://www.biotech-info.net/portends.html Dr Robert Goodman, 'Genomics portends the next revolution in agriculture', University of Wisconsin-Madison press release, 18 February 2001.
7. http://www.gmwatch.org/archive2.asp?ArcId=316 Soni Mishra, 'Genetically engineered rice? Take a look at farmers' varieties', *Hindustan Times* (New Delhi), 12 December 2002.
8. http://www.gmwatch.org/archive2.asp?ArcId=556 ISIS condemns Prime Minister's Scoping Note, 25 October 2002.
9. http://www.gmwatch.org/archive2.asp?arcid=281 John Vidal, 'Famine in Africa: Controlling their own destiny', *Guardian*, 30 November 2002.
10. http://www.gmwatch.org/archive2.asp?ArcId=338 'The Secret of El Dorado', *Horizon*, BBC 2, 19 December 2002.
11. http://www.gmwatch.org/archive2.asp?ArcId=2000 Editorial in *New Scientist*, 3 February 2001.
12. http://www.gmwatch.org/p1temp.asp?pid=3&page=1
13. Rowell, *Don't Worry*, pp. 200–2.
14. http://www.greenpeace.org.uk/Products/GM/supermarkets.cfm
15. Co-op Customer Services line, Tel: 0800 317 827, 5 October 2004.
16. http://www.greenpeace.org.uk/Products/GM/supermarkets.cfm
17. Iceland Press Office, Tel: 01288 842941, 21 January 2005.
18. http://www.greenpeace.org.uk/Products/GM/supermarkets.cfm
19. http://www.organicconsumers.org/supermarket/protests0608.cfm
20. Greenpeace website.
21. *GM-Free*, vol. 1, no. 2, June/July 1999, pp. 19–20.

## CHAPTER 10 A LAST WORD

1. Monbiot, *Captive State*, pp. 357–9.
2. http://www.gmwatch.org/archive2.asp?arcid=4209 Devinder Sharma, 'WTO accord: Faulty frame, rude reality', *The Hindu*, 5 August 2004.
3. http://www.gmwatch.org/archive2.asp?arcid=3423 Alexandra Wandel, Friends of the Earth Europe, Letter to the *Financial Times*, 5 May 2004.
4. http://www.btinternet.com/~nlpwessex/Documents/gmocarto.htm

# Resources

## BOOKS

Cook, Guy, *Genetically Modified Language: The Discourse of Arguments for GM Crops and Food* (Routledge, 2004).

Edwards, David, *Free to be Human: Intellectual Self-Defence in an Age of Illusions* (Green Books, 1995).

Monbiot, George, *Captive State: The Corporate Takeover of Britain* (Pan, 2000).

Paul, Helena and Steinbrecher, Ricarda, *Hungry Corporations: Transnational Biotech Companies Colonise the Food Chain* (Zed Books, 2003).

Pilger, John, *Hidden Agendas* (Vintage, 1998).

Rowell, Andrew, *Green Backlash: Global Subversion of the Environment Movement* (Routledge, 1996).

—— *Don't Worry, It's Safe to Eat: The True Story of GM Food, BSE, and Foot and Mouth* (Earthscan, 2003).

Shanks, Pete, *Human Genetic Engineering: A Guide for Activists, Skeptics, and the Very Perplexed* (Nation Books, 2005).

Smith, Jeffrey M., *Seeds of Deception: Exposing Industry and Government Lies About the Safety of the Genetically Engineered Foods You're Eating* (Green Books, 2004).

—— *Genetic Roulette: The Documented Health Risks of Genetically Engineered Food* (Yes! Books, Iowa, 2006).

Stauber, John and Rampton, Sheldon, *Toxic Sludge is Good for You!: Lies, Damn Lies and the Public Relations Industry* (Common Courage Press, 1995).

## REPORTS
### (IN DATE ORDER)

### GM crop performance

Duffy, Mike and Ernst, Matt, 'Does Planting GMO Seed Boost Farmers' Profits?' *Leopold Center for Sustainable Agriculture Newsletter*, vol. 11, no. 3, Fall 1999. Available from: http://www.leopold.iastate.edu/pubs/nwl/1999/1999–3leoletter.pdf

Oplinger, E.S., Martinka, M.J. and Schmitz, K.A., 'Performance of Transgenetic Soybeans – Northern US', presented to the ASTA Meetings, Chicago (1999).

WWF International report, 'Transgenic Cotton: Are There Benefits for Conservation?', March 2000. Available from: http://www.panda.org/downloads/freshwater/ct_long.pdf

WWF International paper, 'No Reduction of Pesticide Use with Genetically Engineered Cotton', Fall 2000. Available from: http://www.biotech-info.net/WWF_inter_update.pdf

Five Year Freeze, 'Feeding or Fooling the World? Can GM Really Feed the Hungry?', report, October 2000. Available from: http://www.fiveyearfreeze. org/feed fool world.pdf

Moeller, David, 'GMO Liability Threats for Farmers', Institute for Agriculture and Trade Policy (IATP) briefing paper, December 2001. Available from: Ben Lilliston, Institute for Agriculture and Trade Policy, tel: 612–870–3416.

Benbrook, Charles, 'When Does it Pay to Plant Bt Corn: Farm-Level Economic Impacts of Bt Corn, 1996–2001', report, December 2001. Summary and full report available from: http://www.iatp.org

R.W. Elmore et al., 'Glyphosate-Resistant Soybean Cultivar Yields Compared with Sister Lines', *Agronomy Journal*, vol. 93, 2001, pp. 408–12. Available from: http://screc.unl.edu/Research/Glyphosate/glyphosateyield.html

Duffy, Michael, 'Who Benefits from Biotechnology?', January 2002. The text of Duffy's presentation of the results at the December 2001 American Seed Trade Association meeting is available from: http://www.leopold.iastate. edu/pubs/speech/files/120501-who_benefits_from_biotechnology.pdf

US Department of Agriculture, 'The Adoption of Bioengineered Crops', report, June 2002. Available from: http://www.ers.usda.gov/publications/ aer810/. Summary from: www.btinternet.com/~nlpwessex/Documents/ usdagmeconomics.htm

Benbrook, Charles, 'Evidence of the Magnitude and Consequences of the Roundup Ready Soybean Yield Drag from University-Based Varietal Trials in 1998', report, July 2002. Available from: http://www.biotech-info.net/ RR_yield_drag_98.pdf

Soil Association, 'Seeds of Doubt, North American Farmers' Experiences of GM Crops', report, September 2002. Available free from: http://www. non-gm-farmers.com/ Summary at: http://www.gmwatch.org/archive2. asp?ArcId=548

ActionAid, 'GM Crops – Going Against the Grain', report, May 2003.

deGrassi, Aaron, 'Genetically Modified Crops and Sustainable Poverty Alleviaton in Sub-Saharan Africa: An Assessment of Current Evidence', Third World Network Africa, June 2003. Available from: http://allafrica. com/sustainable/resources/view/00010161.pdf

Friends of the Earth and Greenpeace, 'The Impact of GM Corn in Spain', study, August 2003. Available from: http://www.tierra.org and http://www. greenpeace.org/espana_es

Wills, Peter, 'Genetic Engineering Policy and Science Since the Royal Commission: No Resolution of Problems in Sight', September 2003. Summary from: http://www.gmwatch.org/archive2.asp?arcid=1510

Benbrook, Charles, 'Impacts of Genetically Engineered Crops on Pesticide Use in the United States: The First Eight Years', November 2003. Available from: http://www.biotech-info.net/technicalpaper6.html

Chern, W.S., Rickertsen, K., Tssuboi, N. and Fu, T. 'Consumer Acceptance and Willingness to Pay for Genetically Modified Vegetable Oil and Salmon: A Multiple-Country Assessment', *AgBioForum*, vol. 5, no. 3, 2003, pp. 105–112. Available from: http://www.agbioforum.org/v5n3/v5n3a05-chern.htm

Independent Science Panel, 'The Case for a GM-Free Sustainable World', report, published by Institute of Science in Society, 2003. Available from: http://www.indsp.org/

Benbrook, Charles, 'Genetically Engineered Crops and Pesticide Use in the United States: The First Nine Years', October 2004. Full report available from: http://www.biotech-info.net/Full_version_first_nine.pdf Abstract available from: http://www.biotech-info.net/technicalpaper7.html

Sharma, Devinder, 'GM Food and Hunger: A View from the South', Forum for Biotechnology and Food Security, 2004. Orders from: dsharma@del6.vsnl.net.in

### India's GM Bt cotton performance

Andhra Pradesh Department of Agriculture, survey, 2002/3 season: Summary and comment from: http://www.gmwatch.org/archive2.asp?arcid=4922 & http://www.gmwatch.org/print-archive2.asp?arcid=157

Sahai, Suman and Rahman, Shakeelur 'Performance of Bt Cotton in India: Data from the First Commercial Crop', Gene Campaign field study, July 2003. Available from: http://www.genecampaign.org/archive12.html Summary and comment from: http://www.gmwatch.org/archive2.asp?arcid=4922 & http://www.gmwatch.org/archive2.asp?arcid=1427

Deccan Development Society in Andhra Pradesh, report, 2002/3 season, 'Did Bt Cotton Save Farmers in Warangal?'. Available from: www.ddsindia.com Summary and comment from: http://www.gmwatch.org/archive2.asp?arcid=4922 & http://www.gmwatch.org/archive2.asp?arcid=949

Greenpeace, 'Performance of Bt Cotton in Karnataka', study of three districts of Karnataka, April 2003. Comment from: http://www.gmwatch.org/archive2.asp?arcid=762

Qayum, M.A. and Sakkhari, K. 'Performance of Bt Cotton in Andhra Pradesh: 2003–4', study by Andhra Pradesh Coalition in Defence of Diversity (APCIDD), May 2004. Comment and summary at: http://www.gmwatch.org/archive2.asp?arcid=3405

Centre for Sustainable Agriculture, Secunderabad, 'The story of Bt Cotton in Andhra Pradesh: Erratic Processes and Results', February 2005. Available in two parts from: http://www.gmwatch.org/archive2.asp?arcid=4922 http://www.gmwatch.org/archive2.asp?arcid=4923

### GM testing

*Reviews*

Pusztai, Arpad, 'Genetically Modified Foods: Are They a Risk to Human/Animal Health?', June 2001. Available from: http://www.actionbioscience.org/biotech/pusztai.html

Pryme, I.F. and Lembcke, R. '*In Vivo* Studies on Possible Health Consequences of Genetically Modified Food and Feed', *Nutrition and Health*, vol. 17, 2003, pp. 1–8. Available from: http://www.soilassociation.org/web/sa/saweb.nsf/b0062cf005bc02c180256a6b003d987f/80256cad0046ee0c80256d66005ae0fe/$FILE/NutritionHealthstudy.pdf Overview from: http://www.gmwatch.org/archive2.asp?arcid=1131

Dr Arpad Pusztai et al's review, covering all published papers up to the end of 2002, can be found on pp. 347–72 of the book *Food Safety: Contaminants and Toxins*, edited by J.P.F. D'Mello (CABI Publishing, 2003).

A. Pusztai and S. Bardocz's latest review of GM feeding studies can be found on pp. 513–40 of the book, *Biology of Nutrition in Growing Animals*, edited by R. Mosenthin, J. Zentek, and T. Zebrowska (Elsevier Ltd, 2005).

*Individual studies*

Ewen & Pusztai, 10-day study on male rats fed GM potatoes, 1999: S.W.B. Ewen and A. Pusztai, 'Effects of Diets Containing Genetically Modified Potatoes Expressing *Galanthus nivalis* Lectin on Rat Small Intestine', *Lancet*, vol. 354, 1999b, pp. 1353–4. Comment available from: http://www.gmwatch. org/archive2.asp?arcid=915 – Report of the month

Pusztai, 110-day study on male rats fed GM potatoes, 1998: Comment available from: http://www.gmwatch.org/archive2.asp?arcid=915 – Report of the month

Aventis's Chardon LL herbicide-resistant GM maize: Comment available from http://www.gmwatch.org/archive2.asp?arcid=1890 – Item 2

Newcastle feeding study: Netherwood et al., 'Assessing the Survival of Transgenic Plant DNA in the Human Gastrointestinal Tract', *Nature Biotechnology*, vol. 22, 2004, pp. 204–9.

## ONLINE INFORMATION

### Newsletters and other information

The Campaign E-mail lists (US) offer three free anti-GM newsletters ('Campaign Reporter', 'News Updates', and 'More GE News') from: http:// www.thecampaign.org/emaillists.php

GM Watch has two free online newsletters: 'Weekly Watch' (weekly) and 'GM Watch' (monthly). There are also daily bulletins. All available from: http:// www.gmwatch.org/sub.asp. GM Watch also has a website, including the excellent 'Biotech Brigade' directory at: http://www.gmwatch.org/p1temp. asp?pid=1&page=1

nlpwessex has a good website, which is especially useful for farming-related issues. Contains lists of quotes about GM from scientists, farmers, and other public figures. Available from: www.btinternet.com/~nlpwessex/ Documents/gmocarto.htm

'The Non-GMO Report' (monthly from US) is a newsletter aimed at non-GM growers. Available from: http://www.non-gmoreport.com/

Organic Consumers Association (US) has a bimonthly newsletter 'Organic Bytes.' Available from: http://www.organicconsumers.org/organicbytes. htm

'Spilling the Beans' (US) is a free newsletter available once or twice a month from: http://www.seedsofdeception.com/utility/showArticle/ ?objectID=69

### Campaigns

Consumers International: see: http://www.consumersinternational.org/ Templates/News.asp?NodeID=89677&int1stParentNodeID=89650

Friends of the Earth: see: http://www.foe.co.uk/campaigns/real_food/

GM Watch: 'Weekly Watch' has a 'Campaign of the Week'.

Greenpeace UK: see: http://www.greenpeace.org.uk/contentlookup.cfm?SitekeyParam=D-I

Greenpeace International: see: http://www.greenpeace.org/international/campaigns/genetic-engineering

## ANTI-GM NGOs

### UK

ActionAid, Christian Aid, Five Year Freeze, Friends of the Earth, Genetic Engineering Network, Genetics Forum, Genetix Food Alert, GeneWatch, Green Party, Greenpeace, HDRA, Institute of Science in Society (ISIS), National Trust, Natural Law Party Wessex, NGIN/GM Watch, Oxfam, Pesticide Action Network, RSPB, Soil Association, Vegetarian Society, Wildlife Trusts, Women's Environmental Network, Women's Institute, World Development Movement, WWF.

### US

Californians for GE-Free Agriculture, Campaign to Label Genetically Modified Foods, Center for Food Safety, Council for Responsible Genetics, Edmonds Institute, Farmer to Farmer Campaign on Genetic Engineering, Friends of the Earth, GE Food Alert, Genetic Engineering Action Network, Greenpeace, neRAGE Northeast Resistance Against Genetic Engineering, Organic Consumers Association, Rural Advancement Foundation International (RAFI), Sierra Club, True Food Network, Union of Concerned Scientists.

### South Africa

For lots of contacts, see: http://www.gmwatch.org/archive2.asp?arcid=5237

# Index